Advances in Adaptive Optics

Advances in Adaptive Optics

Edited by **Kristie Ames**

New Jersey

Published by Clanrye International,
55 Van Reypen Street,
Jersey City, NJ 07306, USA
www.clanryeinternational.com

Advances in Adaptive Optics
Edited by Kristie Ames

International Standard Book Number: 978-1-63240-045-1 (Hardback)

Printed in the United States of America.

Contents

Permissions

List of Contributors

Preface

This book is designed to provide an ideal introduction to the field of adaptive optics. In the past few years, there has been constant development in adaptive optics technology, theory, and systems growth. Lately, there has been an increase in the applications of adaptive optics in the fields of communications and medicine, in addition to its unique uses in astronomy and beam transmission. This book elucidates solar astronomy, optical vortices, retinal imaging, and many other related topics. It is a compilation of various well researched chapters intended to help its reader in gaining more knowledge regarding this field.

The information shared in this book is based on empirical researches made by veterans in this field of study. The elaborative information provided in this book will help the readers further their scope of knowledge leading to advancements in this field.

Finally, I would like to thank my fellow researchers who gave constructive feedback and my family members who supported me at every step of my research.

Editor

Integrated Adaptive Optics Systems

A Solar Adaptive Optics System

Ren Deqing and Zhu Yongtian

Additional information is available at the end of the chapter

1. Introduction

Solar activities are dominated by magnetic fields, which are arranged in small structure. The structure and evolution of small-size magnetic fields are the key component in a unified understanding of solar activities [1]. As such, a major application of a large solar telescope is for high-sensitivity observations of solar magnetic fields. The observation of solar dynamics of small-scale magnetic fields requires un-compromised high resolution, high magnetic field sensitivity, and high temporal resolution [2, 3]. The two important scales that determine the structuring of the solar atmosphere are the pressure scale height and the photon mean free path, which are of on the order 70 km or 0.1″. Recently, structures as small as a few tens of kilometers on the solar surface corresponding to a few tens of milli-arcseconds on the sky have been predicted by sophisticated MHD models of the solar atmosphere [4-7]. For a ground-based telescope, however, the atmospheric turbulence will seriously degrade the actual performance for high-resolution imaging, and an adaptive optics (AO) system is needed to recover the theoretical diffraction-limited angular resolution in real-time scale [8].

Current major solar telescopes have been equipped with dedicated AO systems that adopt different techniques for real-time wave-front sensing and image signal processing [9]:

1. The AO system with the 0.76-meter Dunn solar telescope uses Digital Signal Processors (DSPs) for the real-time signal processing [10]. DSPs are superb for fast calculation for digital image processing. However it is time-consuming for the DSP programming, and it lacks flexibility.

2. The 0.7-m Vacuum Tower Telescope at Teide Observatory uses multiple Processors (CPUs) on workstation computers and low-level programming language such as C++ for AO programming [11, 12].

The performance of the AO systems with multi-CPUs is close to those with DSPs. However, the low-level C++ programming is also time-consuming. Recent CPU developments indicate that multi-core technique is superior over that of the multi-CPU in view of the calculation speed and power consuming. A detailed review of solar adaptive optics was discussed by Rimmele and Marino [13].

Due to the rapid development of multi-core personal computers and the powerful parallelism of the LabVIEW software, we proposed a novel solar AO system that is based on today's multi-core CPUs and "high-level" LabVIEW programing [14]. The Portable Solar Adaptive Optics (PSAO) system at California State University Northridge (CSUN) is designed to deliver diffraction-limited imaging with 1~2-m class telescopes which will cover the largest solar telescope currently operational. This AO is optimized for a small physical size, so that we can carry it to any available solar telescope as a visiting instrument for scientific observations. We use personal computers with Intel i7 multi-core CPUs for the AO real-time control, and use LabVIEW software for AO programming. LabVIEW, developed by the National Instruments (NI), is based on block diagram programming, which makes it inherently supporting multi-core or multi-thread calculation in parallel. LabVIEW also includes a large number of high-quality existing functions for mathematical operations and image processing, which makes the AO programing extremely efficient and is suitable for the real-time AO programming.

Since 2009, we have built and continually updated our PSAO system in our laboratory [15]. We have initially tested the PSAO with the 0.6-m solar telescope at San Fernando Observatory (SFO) as well as the 1.6-m McMath-Pierce telescope (McMP). In this paper, we will present recent results in the development of the PSAO in the laboratory and the on-site trial observations.

2. Design philosophy

2.1. Optical Design

The PSAO must be able to work with any solar telescopes with different aperture size and focal ratios, although it was initially developed for testing with the 0.6-meter vacuum solar telescope located at the San Fernando Observatory, CSUN. For such an application, the PSAO optics consists of two individual parts. The first part is the fore-optics, while the second part is the main AO optics. The PSAO optics layout is shown in Figure 1. The fore-optics consists of L1, M1, L2 and M2, while the main AO optic consists of the remaining optical components. Where, L1 and L2 are two lenses and M1 and M2 are two fold mirrors. All the optical components are off-the-self parts. The function of the fore-optics is to convert a telescope's focal ratio to f/54, and create an exit pupil at infinite distance. i.e. create a telecentric image at f/54: in Figure 1, the telescope focal plane image IM0 is first collimated by lens L1, which forms a pupil image on the fold mirror M1. The pupil image is located one focal length distance from lens L2. In such a way, lens L2 forms a solar image at IM1 with the exit pupil at infinite. By adjust the focal length ratio between L1 and L2, one can convert the telescope image IM0 to a telecentric image of f/54 at the IM1; the main AO optics is fixed even with different telescopes, except that the wave-front sensor lenslet array (L6 in Figure. 1) can be chosen from a set of lenslet arrays for different telescopes and seeing conditions. In this way, we only need to adjust the fore-optics without any change for the main AO optics, which makes the PSAO suitable with any solar telescope. For example, the 1.5-m McMP telescope, located at the Kitt Peak National Solar Observatory (NSO), has a focal ratio of f/54 at the focal plane. When working with the McMP, both lenses L1 and L2 are identical and have a focal length of 250mm. As shown in the Figure 1, we use two lenses L1 and L2 as the fore-optics which is of a typical telecentric optics design. The whole AO optics uses several flat fold mirrors (M1, M2, M3, M4) to fold the optical path and reduce the overall physical size. The fold mirror M1 in the fore-optics also serves as a tip-tilt mirror (TTM). The output focal plane image after the fore-optics (L2) is collimated by the lens L3, which creates a pupil image with a size of ~ 4.4-mm on the deformable mirror (DM). Please note that the fold mirror M4 also serves as the DM. After the DM, the beam is split as several parts by two beam splitters B1 and B2, which are used for DM wave-front sensing, tip-tilt sensing and focal plane imaging, respectively. Currently, our AO system has its individual optical channels for DM wave-front as well as tip-tilt sensing, respectively. The DM wave-front sensor (WFS) consists of lenses L4, L5, a lenslet array L6 (for clarity, only one lenslet is shown) and a WFS camera, while the tip-tilt sensor (TTS) consists of the lens L8 and the tip-tilt camera only.

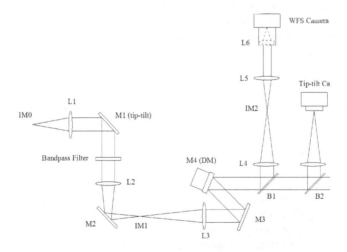

Figure 1. The optical layout of the PSAO.

For the DM WFS channel, lens L4 forms a telecentric solar image IM2, which is collimated subsequently by lens L5. A pupil image is formed one focal length distance behind lens L5, where the lenslet array L6 is located to sample the pupil image for proper wave-front sensing. This is a typical configuration of a Shack-Hartmann wave-front sensor, except that the field of view (FOV) formed by each lenslet must have a suitable size for wave-front sensing with a two-dimensional solar structure. A typical field size for solar wave-front sensing is in a range of 15″ ~ 20″, which is a compromise between the wave-front sensing speed and the sensing accuracy. The TTS is very simple. A lens L8, which is a zoom lens, is used to form a solar image directly on the tip-tilt camera for the calculation of the overall image movement. The corrected image is fed directly to the science camera via the lens L9, which is changeable for different image scales with different telescopes.

The WFS - DM and TTS - TTM channels are controlled by two high performance personal computers to form two individual correction loops, respectively; one is for the DM wave-front correction and the other is for the tip-tilt correction. The DM and TTM are both conjugated on the telescope pupil, which eliminates the pupil wander problem and ensures the AO correction extremely stable. A field stop is placed on the telescope focal plane IM0 or on the solar image plane IM1. An adjustable field stop is also located on the solar image plane IM2 to limit the field size for wave-front sensing. The lenslet array is integrated with the WFS camera directly via the camera's C mount, and can be replaced for different focal lengths, which makes the PSAO suitable for different telescopes and seeing conditions.

A band-pass filter is located just before lens L2. The filter has a band-pass width of ~ 100 nm, which will limit the light energy on the small DM. This is not a problem for solar scientific observations, which only need to work on narrow band in most situations. In fact, a filter wheel can be used for the observations at any band. The size of the AO field of view is controlled by the aperture size of a field stop located on the IM0 (or IM1), which limits the AO field of view as 60″x60″. The field of view for the TTS is also set the same as that of the AO field of view, and is sampled by 60x60 pixels of a "region of interest" of the tip-tilt camera, which results in a sampling scale of 1″/pixel. The primary optical specifications are listed in Table 1.

AO FOV	TTS FOV	WFS FOV	Wavelength range
60″x60″	60″x60″	8″x8″ ~ 30″x30″	0.6 ~ 1.5µm

Table 1. PSAO Optical Specifications.

The use of an individual DM wave-front sensor as well as a tip-tilt sensor has some benefits. In addition to avoid the pupil wander for the wave-front sensing, which will deliver a super stable AO system at different seeing conditions, it will allow the use of small field of view for wave-front sensing at good seeing condition, which can further improve the wave-front sensing sensitivity or accuracy. Since the PSAO has its individual DM wave-front and tip-tilt sensors, the TTS can be used to measure the overall wave-front movement in the large 60″x60″ FOV. As the wave-front tip-tilt component is corrected by the tip-tilt mirror, the DM WFS can use a small FOV, such as 8″x8″, for accurate wave-front sensing. Since each WFS lenslet sub-aperture is sampled by 30x30 pixels in our WFS, a 8″x8″ FOV corresponds to a WFS sampling scale of 0.27″ /pixel, compared to the 1.0″ /pixel sampling scale for a 30″x30″ WFS FOV that may be used in poor seeing conditions. This can significantly improve the wave-front sensing accuracy and thus deliver a better AO performance.

2.2. Wave-Front Sensing

Solar wave-front sensing uses a Shark-Hartmann wave-front sensor. The wave-front gradient or slope vector at each sub-aperture of the lenslet array is solved by the cross-correlation calculation of a two-dimensional pattern over a field of view. The correlation function C(x,y) of a sub-aperture S(x,y) and a reference pattern R(x,y) can be calculated over the two-dimensional sub-aperture as

$$C(x,y) = FFT^{-1}[FFT(S) \cdot (FFT(R))^*] \tag{1}$$

the asterisk denotes the complex conjugate. FFT and FFT^{-1} represent Fourier and inverse Fourier transfers, respectively. The resulting correlation function is "star-like" and can be treated like a star in the stellar WFS. The position of the maximum value of the correlation function corresponds to a slope vector where the reference pattern of R(x,y) best matches the sub-aperture pattern of S(x,y). The calculations of the Fourier and inverse Fourier transfers for the so-called pattern match are time consumed, which makes the high-speed AO correction challenging.

If wave-front phase ϕ is described by the Zernike polynomial expansion as

$$\varphi = \sum_{k=1}^{K} a_k Z_k(x,y) \tag{2}$$

From equation 2, the slope vector s of the WFS and mode coefficient vector a are associated as

$$s = [B]a \tag{3}$$

where,

$$S = (\frac{\partial \varphi}{\partial x}\Big|_1 \dots \frac{\partial \varphi}{\partial x}\Big|_M \quad \frac{\partial \varphi}{\partial y}\Big|_1 \dots \frac{\partial \varphi}{\partial y}\Big|_M)^T \tag{4}$$

$$a = (a_1 \quad a_2 \dots a_k)^T \tag{5}$$

while the matrix [B] is

$$[B] = \begin{pmatrix} \left.\dfrac{\partial Z_1(x,y)}{\partial x}\right|_1 & \left.\dfrac{\partial Z_2(x,y)}{\partial x}\right|_1 \cdots \left.\dfrac{\partial Z_K(x,y)}{\partial x}\right|_1 \\ \vdots & \\ \left.\dfrac{\partial Z_1(x,y)}{\partial x}\right|_M & \left.\dfrac{\partial Z_2(x,y)}{\partial x}\right|_M \cdots \left.\dfrac{\partial Z_1(x,y)}{\partial x}\right|_M \\ \left.\dfrac{\partial Z_1(x,y)}{\partial y}\right|_1 & \left.\dfrac{\partial Z_2(x,y)}{\partial y}\right|_1 \cdots \left.\dfrac{\partial Z_K(x,y)}{\partial y}\right|_1 \\ \vdots & \\ \left.\dfrac{\partial Z_1(x,y)}{\partial y}\right|_M & \left.\dfrac{\partial Z_2(x,y)}{\partial y}\right|_M \cdots \left.\dfrac{\partial Z_1(x,y)}{\partial y}\right|_M \end{pmatrix} \tag{6}$$

Here, M is the number of WFS sub-apertures. K is the number of Zernike modes. The mode coefficient vector is found by finding the pseudo-inverse of [B], which is solved by using the singular value decomposition (SVD) as,

$$[B] = UDV^T \tag{7}$$

and,

$$[B]^{-1} = VD^{-1}U^T \tag{8}$$

Once the DM's influence function is known, the measured wave-front is used to find the DM' signals (i.e. voltages) that are required to correct the wave-front error. The Zernike polynomials act as a low-band pass filter, which is used to measure and control the actual wave-front error up to the mode number K. The choice of actual mode number that an AO system can correct should consider the system's stability, which can be determined by the condition number as discussed by Kasper et al. [16].

2.3. Electrics and Programming

All the PSAO's hardware is based on off-the-shelf commercial components, which makes a low cost system possible. The performance of our AO system can continue to improve once better components are available on the market. The current AO WFS loop uses a computer equipped with a first-generation Intel i7 -990X CPU, which has 6 cores and 12 threads for parallel computation. This computer can be updated to a second-generation Intel i7 CPU that should deliver a better performance, or even updated to a computer with two recent Xeon CPUs that will have 16 cores and 32 threads in total, which is expected to be two times faster than the current system. The specifications of current hardware components are listed in the Table 2.

Hardware	Specifications
DM	140 actuators, 3.5μm stroke, 14-bit resolution, 4.4-mm clear aperture, 8000 Hz frame rate.
TTM	PI S-330.4SL, 5mrad stroke range, 0.25μrad resolution, 1600 Hz frame rate.
WFS & Tip-tilt Camera	1024x1024 pixels, 10.6μm pixel size, 150Hz frame rate at full resolution.
Image Grabbers	NI PCIe-1429 Camera-Link image grabber.
WFS lenslet arrays	0.3mm pitch, 4.7mm, 8.7mm, 18.8mm focal lengths.
Computer 1 (for WFS loop)	Intel Core i7-990X @ 3.47GHz, 8GB RAM.
Computer 2 (for TTS loop)	Intel Core i7-980X @ 3.33GHz, 4GB RAM.

Table 2. PSAO Hardware specifications.

The DM and TTM are two critical components for the AO system. We use the high-speed Multi-DM from Boston Micronmachines Corporation (BMC), which is a micro-electro-mechanical-systems (MEMS) deformable mirror and has 140 actuators (in a 12x12 array without 4 corners) with 3.5μm stroke and can deliver a frame rate up to 8000 Hz. The DM has a clear aperture of 4.4 mm only. Although this allows for a small beam size and makes the whole AO system smaller in physical size, a DM with a clear aperture on the order of 8~10mm will be preferred for our AO system, which will not significantly increase the physical size and will be more robust to the alignment error between the DM and the WFS, such as the error introduced by the vibration from where the AO is located. If the DM had an 8-mm clear aperture, for example, the focal ratio at IM1 in Figure 1 could be f/27, instead of f/54, and in this case lens L3 will has the same focal length with that for the 4.4-mm DM, and thus will not increase the overall AO physical size. The TTM is a flat mirror mounted on a S-330.4SL piezo-tilt platform from Physik Instrumente (PI), which has 5-mrad stroke and can deliver an actual frame rate of 700 Hz only with the digital USB input port, although the datasheet claims that it has a unloaded resonant frequency of 3.3 kHz, and a resonant frequency of 1.6 kHz loaded with a 25 x 8 mm glass mirror. The strokes of our DM and TTM are both sufficient for the AO requirements. In theory, the WFS can simultaneously measure tip-tilt and high-order wave-front errors so that the DM and TTM can be controlled by one computer only. However the current TTM maximum frequency is too slower, comparing to that of the DM, which will reduce the overall AO correction speed, if both were controlled by a single closed-loop. In order to keeping the DM correction fast, we split the DM wave-front and tip-tilt corrections as two individual close-loops, and use two computers to control the DM and TTM, separately.

Both the DM wave-front sensor and the tip-tilt sensor adopt a high-speed MV-D1024E-CL160 camera made by Photonfocus, respectively. The camera transfers image data via the base camera-link interface at a speed of 255MB/s. The output data from each camera is acquired by a high-performance NI PCIe-1429 Camera-Link image grabber connected to the associated controlling computer via a PCIe slot. We chose the NI image grabbers for both WFS and TTS cameras, since many existing LabVIEW standard functions for image acquisition are supported by this grabber. The NI PCIe-1429 image grabber supports full, medium, and base camera-link interfaces. The camera can achieve a rate up to 150 frames/second at full-resolution with 1024x1024 pixels. Since we only use a small region of interest with ~ 300x300 pixels for wave-front sensing, the acquisition speed can achieve 1800 frames/s. For the TTS, we only use 60x60 pixels. In a future update, we schedule to use a full camera-link camera, which should be able to deliver an image data acquisition at a speed of ~800 MB/s.

The most time-consuming part in the PSAO is the wave-front sensing and the calculation of control signals for the DM wave-front correction, which must be executed with a high performance computer. The computer used for DM wave-front correction loop has an Intel Core i7-990X CPU with 6 cores, at a clock frequency of 3.47GHz. The computer used for the tip-tilt correction loop has an Intel Core i7-980X CPU with 6 cores, at a clock frequency of 3.33GHz. Here we choose the computer with a single CPU with multiple cores other than the multiple CPUs, since these multiple CPUs are mainly optimized for large data handling such as for internet servers, and are not optimized for real-time calculation and control. This problem will be solved by the latest Intel Xeon E5-2600 series processors which adopt the same architecture with the Core i7 processor, and have up to 8 cores per CPU. A computer equipped with 2 Intel Xeon E5-2687W CPUs for the DM wave-front correction loop is expected to increase the closed-loop bandwidth of our AO system better than 100 Hz, which will be on the state-of-the-art of the current solar AO systems.

Our AO software is written in LabVIEW codes. LabVIEW's Graphical programs inherently contain information about which parts of the code should execute in parallel. Parallelism is important in AO programing because it can unlock performance gains relative to purely sequential programs, due to recent changes in computer processor designs, in which CPUs are moved to multi-cores. LabVIEW also has a large number of standard functions for image processing, mathematical operations and hardware control. These high-quality functions are optimized for high-speed calculation as well as real-time control. For example, the standard function of pattern match in LabVIEW is about 6 ~ 9 times faster than the conventional cross-correlation algorithm. To fully take advantage of the power of today's multi-core CPU and high-quality LabVIEW's graphic programming, we use LabVIEW's parallel programming, which makes rapid development of a high-performance AO system possible. LabVIEW has greater flexibility and capability for real-time hardware system control than other general-purpose programming languages. LabVIEW programming is performed by wiring together graphical icons on a diagram, in which each icon is a built-in function. This makes programming extremely easy and efficient. In addition, dataflow languages like LabVIEW allow for automatic parallelization. Graphical programs inherently contain information about which parts of the code should execute in parallel. In the future, our system can also be easily updated to a full field-programmable gate array (FPGA) system by using NI's PXI system that is fully supported by NI's LabVIEW parallel programming, which may further increase the AO correction speed. Historically, FPGA programming was the province of only a specially trained expert with a deep understanding of digital hardware design languages. LabVIEW's FPGA programming makes it possible for engineers without FPGA expertise to use it, and makes the software development efficient.

The AO software is composed of two parts. The first part is for the AO calibration, which automatically searches for all effective WFS sub-apertures, calculates DM influence function, and record all the calibration data; the second part is for AO real-time correction, which first reads the calibration data and then performs wave-front correction in real-time. Due to the intrinsic support for parallel processing, LabVIEW automatically assigns the calculation tasks as multiple threads for each core, so that the program can be run in parallel, which greatly increases the running speed for the AO wave-front sensing and correction.

2.4. Tip-Tilt and Deformable Mirror Requirements

Since the atmospheric turbulence is corrected by the tip-tilt and deformable mirrors, the strokes provided by the tip-tilt or deformable mirror must be sufficiently large, so that the wave-front errors can be effectively corrected. Here, we use the 1.6-meter McMP as an example to calculate the tip-tilt and DM requirements. The total minimum stroke required for the tip-tilt mirror is given by [17]

$$Stroke_{min} = 1.25 \frac{D}{D_{tilt}} \sqrt{0.184(\frac{D}{r_0})^{5/3}(\frac{\lambda}{D})^2} \qquad (9)$$

where, D is the telescope aperture size and is equal to 1.6 meters for the McMP. D_{tilt} is the diameter of the telescope aperture de-magnified on the tilt mirror. Assume that the telescope aperture is de-magnified as ~5.0 mm on the tip-tilt mirror, D_{tilt} is equal to 5.0 mm. At 1.25-μm wavelength, r_0 is equal to 150 mm (see next subsection). This results in a minimum stroke of 0.001 radians. The tip-tilt mirror will be mounted on a S-330.4SL piezo-tilt platform that can provide a maximum tilt of 0.005 radians that is larger than the minimum stroke requirement.

Similarly, the required stroke for the deformable mirror is calculated as [17]

$$Stroke \ (waves) = 2.5 \sqrt{0.00357(\frac{D}{r_0})^{5/3}} \qquad (10)$$

This results in a required stroke of 1.08 waves at 1.25-μm wavelength. The deformable mirror from BMC can provide a maximum stroke of 3.5 μm that is larger than the required stroke. BMC's deformable mirrors are being used for astronomic adaptive optics systems, where large amount of actuators or a compact design is required. For example, the "Extreme Adaptive Optics" will use a BMC deformable mirror for the 8-meter Gemini telescope, where a deformable mirror with 4096 actuators is required [18]. The high-speed MBC Multi-DM was also used for stellar adaptive optics with a small physical size [19]. The precision and stability of the BMC's deformable mirror have been proved by our past experiences.

2.5. Performance Estimation

As an example, the AO performance estimation will only focus on the Kitt Peak 1.6-meter McMP that has a large aperture size. The estimated performance should be better for other telescope with a smaller aperture size or a site with a better seeing condition. The Fried parameter r_0 is equal to ~ 5 cm at 0.5 μm for the daytime median seeing conditions at Kitt Peak, which is available almost every week for a clear sky. The seeing r_0 is scalable with wavelength as $r_0 \propto \lambda^{\frac{6}{5}}$ for Kolmogorov turbulence, and it reaches 70 mm and 150 mm at the 0.65 μm and 1.25 μm wavelengths, respectively.

A DM surface with a finite actuator number cannot exactly match the wave-front patterns of the atmospheric turbulence. For an atmospheric wave-front with Kolmogorov spectrum, the fitting error variance of a DM with finite actuator number is derived by Hudgin [20] and is given as

$$\sigma_{fit}^2 = \kappa (r_s / r_0)^{5/3} \tag{11}$$

where r_s and r_0 are the spaces between two actuators and the Fried parameter, respectively. κ is the fitting parameter. Since there are 12 actuators across the DM aperture that is conjugated onto the 1.0-meter effective telescope aperture (see Section 4) although McMP has a 1.6-meter aperture, r_s is equal to 83 mm. An extensive analysis of the fitting error showed that $\kappa = 0.349$ was applicable for many influence functions that are not constrained at the edge [21]. Therefore, the fitting error variance σ_{fit}^2 is 0.46 radians^2 and 0.25 radians^2 at the 0.65 μm and 1.25 μm wavelengths, respectively.

The temporal error of the wave-front correction is determined by the Greenwood frequency and the bandwidth of the AO system. The Greenwood frequency can be calculated as $0.43 v / r_0$ [22], where v is the average wind speed. At McMP, when wind speed reaches 20 m/s, the telescope will be closed and observations will not take place. We assume that the AO system will operate with an average wind speed of 8.0 m/s. This results in a Greenwood frequency of 49 Hz and 23 Hz at the 0.65 μm and 1.25 μm wavelengths, respectively. The wave-front variance due to the temporal error of the correction can be calculated as $\sigma_{tem}^2 = (f_G / f_{BW})^{5/3}$, where f_G is the Greenwood frequency and f_{BW} is the bandwidth of the AO system. Since the AO closed-loop bandwidth is 80 Hz, the temporal wave-front variance σ_{tem}^2 is 0.44 radian^2 and 0.13 radian^2 at the 0.65 μm and 1.25 μm wavelengths, respectively.

Read out noise is not a problem for solar wave-front sensing since plenty of photons are available for the wave-front sensing [10]. The corrected wave-front variance is the sum of all the error contributors. If only the fitting and temporal errors are considered, the wave-front variance can be calculated approximately as $\sigma^2 \approx \sigma_{fit}^2 + \sigma_{tem}^2$. This results in a wave-front variance of 0.90 radian^2 and 0.38 radian^2 at the 0.65 μm and 1.25 μm wavelengths, respectively. The overall performance of an AO system can be evaluated in terms of the Strehl ratio, which defines the peak of the actual point spread function normalized to the peak of the diffraction-limited point spread function. The Strehl ratio is calculated as $S = e^{-\sigma^2}$. At 0.65 μm, the AO system is expected to achieve a Strehl ratio of ~ 0.41; at 1.25 μm, it will deliver a Strehl ratio of ~ 0.68, and should deliver a better correction at longer wavelengths.

3. Recent laboratory tests

The PSAO was first built in CSUN laboratory in 2009, with an OKO 37-actuator DM for software development and test purpose. In 2010, the DM was upgraded with the current BMC 140-actuator model. A cross target illuminated by a white-light fiber bundle is used for the 2-dimensional object test. A 32-actuator DM from Edmund Optics is used to generate a real-time wave-front aberration at a desired frequency, which is subsequently fed into the AO system, so that the AO performance can be evaluated in real-time. The whole optics of PSAO is built on a 900mm x 600mm optical breadboard as shown in the Figure 5, and can be easily carried to any solar observatory.

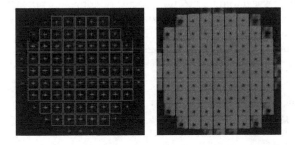

Figure 2. Graphic WFS interfaces for AO calibration with a point source (left) and for real-time correction with a cross target (right).

The AO software consists of two parts. The first part is used for the AO calibration, while the second part is the code for AO real-time correction. In principle, the calibration only needs to be done once a time, provided that the hardware has not been realigned. In the calibration, a single-mode fiber is served as a point source. Any possible incoming wave-front error will be filtered out by the single-mode fiber, and only the fundamental mode can propagate through the fiber. The output wave-front can be viewed as a perfect wave for the calibration. In the calibration, the fiber is switched into the optical path on the telescope focal plane just before the lens L1 (see Figure 1). In AO real-time correction, the fiber is switched out the optical path, and an extended target (a cross herein), including the wave-front error generated by the 32-element Edmund Optics DM, is allowed to input into the AO for testing. A lenslet array with 8.7-mm focal length is used for the WFS. Figure 2 shows the WFS interfaces in the AO calibration with a point source (Figure 2 left) and in real-time correction with a cross target (Figure 2 right): there are 69 effective WFS sub-apertures arranged in the

9x9-lenslet grid configuration and wave-front correction up to 65 modes of Zernike polynomials can be chosen. Please note that in this test, there is no central obstruction on the pupil, although our software can be used with a telescope with central obstruction.

Figure 3. Point-source images: original image without aberration applied (left), with aberration applied and AO off (center), with aberration applied and AO on (right). 9x9 lenslets (without those on the four corners) are used for wave-front sensing, and 25 Zernike modes are corrected.

The PSAO had achieved excellent corrections in the test with a point source target. Figure 3 shows the original point-source image (left) without aberration applied, as well as the AO-off (center) and AO-on (right) images with the aberration applied, respectively. In this test, the wave-front error was introduced by the 32-actuator Edmund Optics DM in real-time, in which the wave-front is variable as a sinusoidal function with a frequency up to 80Hz. The central image of the distorted point-spread function shows that the wave-front error applied is significantly large, while the right image clearly demonstrates that the AO system can recover the diffraction-limited image. The AO system demonstrated good results in different situations:

1.   the amplitude of the applied wave-front error is less or equal to the AO DM's maximum stroke (see Figure 3 and Figure 4). In this situation, the AO system can almost completely correct the wave-front and deliver the same image quality as that there is no wave-front error (i.e. the wave-front error is not been applied);

2.   wave-front error with amplitude larger than the AO DM's maximum stroke is applied, which simulates the bad seeing condition on a site. The AO system can still compensate part of the wave-front error, which is consistent to the DM's maximum stroke. Therefore, the image quality can still be improved.

Figure 4. Cross-target images: original image without aberration applied (left), with aberration applied and AO off (center), with aberration applied and AO on (right). 9x9 lenslets (without those on the four corners) are used for wave-front sensing, and 25 Zernike modes are corrected.

The AO system was also tested with a 2-dimensional extended target with the same procedure for the point-source target. In this test, a cross was printed on a transmission film and was used as a 2-dimensional target. A fiber bundle light source located immediately behind the cross target was used to illuminate it, so that the image of the fibers was almost overlapped on the cross target, except for a slight defocus between them. Figure 4 shows the original cross-target image (left) without aberration applied, as well as the AO-off (center) and AO-on (right) images with the aberration applied, respectively. Please note that the image of the cross target and background small fibers is seriously blurred by the wave-front error applied (center image). Compared the left and right images, however, it is clear that the AO recovers the original image perfectly, indicating that the AO's performance is excellent. In fact, our AO system can recover the original wave-front that is associated with the original image with accuracy up to 1/1000 wave-length in the visible [23].

4. Recent on-site observations

The PSAO's small size makes it can be easily brought to any observatory for science observations. Since 2010, we have carried out observations at two different sites. To demonstrate the AO's feasibility, an initial on-site observation was conducted by using the 0.61-m solar telescope at the San Fernando Observatory, California State University Northridge. This solar telescope is a three-aspheric mirror system with a central obstruction area of 14%, and a focal ratio of f/20. Because of the poor seeing conditions at the San Fernando Observatory, WFS with 9x9 sub-apertures (exception those in regions of the four-corners and the central obstruction) was used. The best observational results were acquired in October 2011. Using a sunspot as a target, the AO system was able to lock on the sunspot for wave-front sensing and provides high-resolution images at the wavelength of 0.75 µm [24], which indicates that our PSAO is able to provide wave-front correction at a site with a poor seeing condition.

After the successful observations on the San Fernando Observatory, we continued to test this system with the 1.6-m McMP. The McMP is located at the Kitt Peak, and is operated by the National Solar Observatory. It is one of the largest solar telescopes and is accessible to the solar community around the world. The medium seeing at the Kitt Peak is better than that at the SFO, but is still poor with a Fried parameter of ~ 5 cm at median seeing condition. The poor seeing condition at the Kitt Peak is a great challenge for an adaptive optics system. There are no AO available for routine observations with the McMP, although a prototype with 36-actuator DM was developed many years ago [25]. McMP is an off-axis telescope without central obstruction. The solar image is formed on a rotational station at f/54, where our PSAO can be conveniently placed for observations. Figure 5 shows the PSAO loaded on the McMP rotation station for an observation, in which the small size of the AO system is clearly referred. The non-common optical path error between the WFS and the science camera, which the WFS cannot measure, was calibrated by an approach we proposed recently [23].

During the latest observations in May 2012, the PSAO delivered excellent performance in the visible with the McMP. Solar images were captured at 0.6-µm visible wavelength. Figure 6 shows the typical images of Sunspot 1492 with the AO off and AO on respectively, on the May 28 run. For the AO off images, the AO still provides the tip-tilt correction, so that the overall image movement is corrected. Compared with the poor image quality when the AO is switched off, the AO system provides significant improvement for the image quality when the AO is switched on, which demonstrates the power of the AO correction: the granules around the sunspot can be clearly seen with the AO correction, while they are totally blurred and disappeared without the correction. The AO off image clearly shows how poor the seeing condition was during our observation run. In the observation, only 7x7 sub-apertures were used for the WFS, and only 1.0-meter of the McMP aperture was used for imaging, because a small area on the edge of the telescope primary mirror was damaged and the telescope heliostat was tilted at a large angle during the observation, which delivered an useful circle aperture on the order of 1.0-meter in diameter. Each sub-aperture was sampled by 30x30 pixels of the WFS camera. The AO delivered an open-loop bandwidth of 800 Hz, which corresponds to a closed-loop bandwidth of ~ 80 Hz. The improvement of image performance with the AO correction was significant. This was the first time demonstration that an AO system can be effectively used for high-resolution imaging in the visible with the McMP.

Figure 5. PSAO setup (the black breadboard) on the McMP rotation station. The two red cameras are used for WFS and TTS, while the grey one is the science camera.

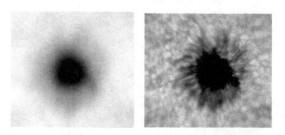

Figure 6. Sunspot 1492 image captured on the McMP with the AO off (left) and AO on (right).

Because of the poor seeing conditions at the Kitt Peak, only large sunspots can be used for wave-front sensing. Other small fine structures such as solar granules and pores are seriously distorted by the strong atmospheric turbulence and cannot be resolved by the WFS, which prevent accurate wave-front sensing by using a small WFS field of view. In this observation, lenslet array with 8.7-mm focal length was used for the WFS. The WFS field of view is 30"x30" and is sampled by 30x30 pixels, which results in a sampling scale of 1.0"/pixels. Although the wave-front sensing of the AO software can execute with sub-pixel accuracy, such an improvement is very limited because of a number of reasons, such as the distortions of the sunspot images in each sub-aperture as well as the low contrast image resulted from the strong wave-front error. Better performance should be achievable with telescopes with good seeing conditions, where small fine solar structures can be used for accurate wave-front sensing.

4. Conclusions

We have fully demonstrated the feasibility of a portable AO system, both in the laboratory and on-site observations. The system is able to provide a wave-front correction with different telescopes with the aperture size up to 1.6 meters. Our AO system features low cost, high-performance, and is compact. Combining the multi-core computer and LabVIEW parallel programming, the AO system is particularly flexible and can achieve good performance. The open-loop correction speed can achieve 800Hz with sub-pixel accuracy for wave-front sensing, for the 7x7 sub-aperture WFS when 25 modes of Zernike polynomials of the wave-front are corrected. It can further achieve 1100 Hz, if sub-pixel wave-front sensing accuracy is not required. Higher wave-front correction speed should be able to achieve by using more CPU cores with a computer. The commercial CPU market for personal computers is being evolved rapidly, with efforts focusing on multi-core CPUs. For example, two Eight-Core Intel Xeon E5-2687W CPUs can be installed in a computer, which will deliver 16 cores in total and each core can run at 3.1GHz clock frequency. In another approach, we are also developing LabVIEW based FPGA technique, which may dramatically increase the running speed of the AO system. The 12x12-actuator DM is also being updated to a 24x24-actuator DM that will have a clear aperture of 9.0 mm, and should deliver better performance. The PSAO is being upgraded accordingly, and we will report our progresses in the near future.

Acknowledgements

This work is supported by the National Science Foundation under the grant ATM-0841440, the National Natural Science Foundation of China (NSFC) (Grant 10873024 and 11003031), the National Astronomical Observatories' Special Fund for Astronomy-2009, as well as the Strategic Priority Research Program of the Chinese Academy of Sciences (Grant No. XDA04070600). We thank Dr. Xi Zhang for his contribution to this AO project, and we gratefully acknowledge the assistances from the staff at Kitt Peak National Solar Observatory during our observations with the McMP.

Author details

Ren Deqing[1,2,3*] and Zhu Yongtian[2,3]

*Address all correspondence to: ren.deqing@csun.edu

1 Physics & Astronomy Department, California State University Northridge, USA

2 National Astronomical Observatories/Nanjing Institute of Astronomical Optics & Technology, Chinese Academy of Sciences, China

3 Key Laboratory of Astronomical Optics & Technology, National Astronomical Observatories, Chinese Academy of Sciences, China

References

[1] Stenflo, Jan. Olof. (2004). Solar physics: Hidden magnetism. *Nature*, 430(6997), 304 -305.

[2] Paletou, F., & Aulanier, G. (2003). On the Need of High-Resolution Spectropolarimetric Observations of Prominences . in *Current Theoretical Models and Future High Resolution Solar Observations: Preparing for ATST, ASP Conference Series286, held March 2002 at NSO, Sunspot, New Mexico, USA, Alexei A. Pevtsov and Han Uitenbroek (eds.), (NSO, Sunspot, New Mexico, USA)*, 11-15.

[3] Lin, H. (2003). ATST near-IR spectropolarimeter. *Proc. SPIE*, 4853, 215-222.

[4] Cattaneo, F., Emonet, T., & Weiss, N. (2003). On the Interaction between Convection and Magnetic Fields. *ApJ*, 588, 1183-1198.

[5] Vögler, A., & Schüssler, M. (2007). A solar surface dynamo. *Astron. Astrophys*, 465, L43-L46.

[6] Nordlund, Á., Stein, R. F., & Asplund, M. (2009). Solar Surface Convection. *Living Rev.Solar Phys*, 6, 2.

[7] Nordlund, Á., & Stein, R. F. (2009). Accurate Radiation Hydrodynamics and MHD Modeling of 3-D Stellar Atmospheres. in *Recent Directions in Astrophysical Quantitative Spectroscopy and Radiation Hydrodynamics, Proceedings of the International Conference in Honor of Dimitri Mihalas for His Lifetime Scientific Contributions on the Occasion of His 70th Birthday, Boulder, Colorado, March 30- April 3, , (Eds.) Hubeny, I., Stone, J.M., Mac-*

Gregor, K., Werner, K., of AIP Conf.Proc.American Institute of Physics, Melville, NY, 1171, 242-259.

[8] Woger, F., von der Luhe, O., & Reardon, K. (2008). Speckle interferometry with adaptive optics corrected solar data. *Astron. Astrophys*, 488, 375-381.

[9] Rimmele, T. R. (2004). Recent advances in solar adaptive optics. *Proc.SPIE*, 5490, 34-46.

[10] Rimmele, T. R., Richards, K., Hegwer, S., Ren, D., Fletcher, S., & Gregory, S. (2003). Solar adaptive optics: A progress report. *Proc. SPIE*, 4839, 635-646.

[11] van der Luhe, O., Soltau, D., Berkefeld, T., & Schelenz, T. (2003). KAOS: Adaptive optics system for the Vacuum Tower Telescope at Teide Observatory . *Proc. SPIE*, 4853, 187-193.

[12] Berkefeld, T., et al. (2010). Adaptive optics development at the German solar telescopes. *Applied Optics*, 49(31), G155 -G166 .

[13] Rimmele, T. R., & Marino, J. (2011). "Solar Adaptive Optics". *Living Reviews in Solar Physics*, 8(2), 1 -92 .

[14] Ren, D., Penn, M., Wang, H., Chapman, G., & Plymate, C. (2009). A Portable Solar Adaptive Optics System. *Proc. of SPIE*, 74380-1.

[15] Ren, D., Penn, M., Plymate, C., Wang, H., Zhang, X., Dong, B., Brown, N., & Denio, A. (2010). A portable solar adaptive optics system: software and laboratory developments. *Proc. SPIE*, 77363-7.

[16] Kasper, K., et al. (2000). ALFA: adaptive optics for the CALAR ALTO Observatory optics, control systems, and performance. *Experience Astronomy*, 10, 49-73.

[17] Tyson, R. K. (2000). Introduction to adaptive optics. *SPIE Press, Washington, USA*.

[18] Macintosh, B., et al. (2006). The Gemini Planet Imager Proc. *SPIEL*, 6272-62720L.

[19] Morzinski, K., Johnson, L. C., Gavel, D. T., Grigsby, B., Dillon, D., Reinig, M., & Macintosh, B. A. (2010). Performance of MEMS-based visible-light adaptive optics at Lick Observatory: closed- and open-loop control . *Proc. SPIE*, 7736-77361O.

[20] Hudgin, R. H. (1977). Wave-front compensation error due to finite correction-element size,. *J. Opt. Am*, 67, 393-395.

[21] Tyson, R. K., Crawford, D. P., & Morgan, R. J. (1990). Adaptive optics system considerations for ground-to-space propagation. *Proc.SPIE*, 1221, 146-156.

[22] Tyson, R. K. (1998). Principles of Adaptive Optics 2th edition. *Academic Press*.

[23] Ren, D., Dong, B., Zhu, Y., & Damian, J. C. (2012). Correction of non-common path error for extreme adaptive optics. *PASP*, 124, 247-253.

[24] Ren, D., & Dong, B. (2012). Demonstration of portable solar adaptive optics system. *Optical Engineering*, 51, 101705 -4.

[25] Keller, C. U., Plymate, C., & Ammons, S. M. (2003). Low-cost solar adaptive optics in the infrared. *Proc. SPIE*, 4853, 351-359.

Chapter 2

Dual Conjugate Adaptive Optics Prototype for Wide Field High Resolution Retinal Imaging

Zoran Popovic, Jörgen Thaung, Per Knutsson and
Mette Owner-Petersen

Additional information is available at the end of the chapter

1. Introduction

Retinal imaging is limited due to optical aberrations caused by imperfections in the optical media of the eye. Consequently, diffraction limited retinal imaging can be achieved if optical aberrations in the eye are measured and corrected. Information about retinal pathology and structure on a cellular level is thus not available in a clinical setting but only from histological studies of excised retinal tissue. In addition to limitations such as tissue shrinkage and distortion, the main limitation of histological preparations is that longitudinal studies of disease progression and/or results of medical treatment are not possible.

Adaptive optics (AO) is the science, technology and art of capturing diffraction-limited images in adverse circumstances that would normally lead to strongly degraded image quality and loss of resolution. In non-military applications, it was first proposed and implemented in astronomy [1]. AO technology has since been applied in many disciplines, including vision science, where retinal features down to a few microns can be resolved by correcting the aberrations of ocular optics. As the focus of this chapter is on AO retinal imaging, we will focus our description to this particular field.

The general principle of AO is to measure the aberrations introduced by the media between an object of interest and its image with a wavefront sensor, analyze the measurements, and calculate a correction with a control computer. The corrections are applied to a deformable mirror (DM) positioned in the optical path between the object and its image, thereby enabling high-resolution imaging of the object.

Modern telescopes with integrated AO systems employ the laser guide star technique [2] to create an artificial reference object above the earth's atmosphere. Analogously, the vast majority of present-day vision research AO systems employ a single point source on the retina as a reference object for aberration measurements, consequently termed guide star (GS). AO correction is accomplished with a single DM in a plane conjugated to the pupil plane. An AO system with one GS and one DM will henceforth be referred to as single-conjugate AO (SCAO) system. Aberrations in such a system are measured for a single field angle and correction is uniformly applied over the entire field of view (FOV). Since the eye's optical aberrations are dependent on the field angle this will result in a small corrected FOV of approximately 2 degrees [3]. The property of non-uniformity is shared by most optical aberrations such as e.g. the well known primary aberrations of coma, astigmatism, field curvature and distortion.

A method to deal with this limitation of SCAO was first proposed by Dicke [4] and later developed by Beckers [5]. The proposed method is known as multiconjugate AO (MCAO) and uses multiple DMs conjugated to separate turbulent layers of the atmosphere and several GS to increase the corrected FOV. In theory, correcting (in reverse order) for each turbulent layer could yield diffraction limited performance over the entire FOV. However, as is the case for both the atmosphere and the eye, aberrations do not originate solely from a discrete set of thin layers but from a distributed volume. By measuring aberrations in different angular directions using several GSs and correcting aberrations in several layers of the eye using multiple DMs (at least two), it is possible to correct aberrations over a larger FOV than compared to SCAO.

The concept of MCAO for astronomy has been the studied extensively [6-12], a number of experimental papers have also been published [13-16], and on-sky experiments have recently been launched [17]. However, MCAO for the eye is just emerging, with only a few published theoretical papers [3, 18-21]. Our group recently published the first experimental study [21] and practical application [22] of this technique in the eye, implementing a laboratory demonstrator comprising multiple GSs and two DMs, consequently termed dual-conjugate adaptive optics (DCAO). It enables imaging of retinal features down to a few microns, such as retinal cone photoreceptors and capillaries [22], the smallest blood vessels in the retina, over an imaging area of approximately 7 x 7 deg^2. It is unique in its ability to acquire single images over a retinal area that is up to 50 times larger than most other research based flood illumination AO instruments, thus potentially allowing for clinical use.

A second-generation Proof-of-Concept (PoC) prototype based on the DCAO laboratory demonstrator is currently under construction and features several improvements. Most significant among those are changing the order in which DM corrections are imposed and the implementation of a novel concept for multiple GS creation (patent pending).

2. Brief anatomical description of the eye

The human eye can be divided into an optical part and a sensory part. Much like a photographic lens relays light to an image plane in a camera, the optics of the eye consisting of the cornea, the pupil, and the lens, project light from the outside world to the sensory retina (Fig. 1, left). The amount of light that enters the eye is controlled by pupil constriction and dilation. The human retina is a layered structure approximately 250 µm thick [23, 24], with a variety of neurons arranged in layers and interconnected with synapses (Fig. 1, right).

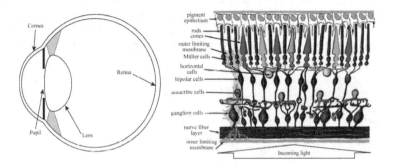

Figure 1. Schematic drawings of the eye (left) and the layered retinal structure (right). (Webvision, http://webvision.med.utah.edu/book/part-i-foundations/simple-anatomy-of-the-retina/)

Visual input is transformed in the retina to electrical signals that are transmitted via the optic nerve to the visual cortex in the brain. This process begins with the absorption of photons in the retinal photoreceptors, situated at the back of the retina, which stimulate several interneurons that in turn relay signals to the output neurons, the retinal ganglion cells. The ganglion cell nerve fiber axons exit the eye through the optic nerve head (blind spot).

Unlike the regularly spaced pixels of equal size in a CCD chip the retinal photoreceptor mosaic is an inhomogeneous distribution of cone and rod photoreceptors of various sizes. The central retina is cone-dominated with a cone density peak at the fovea, the most central part of the retina responsible for sharp vision, with a decrease in density towards the rod-dominated periphery. Cones are used for color and photopic (day) vision and rods are used for scotopic (night) vision.

Blood is supplied to the retina through the choroidal and retinal blood vessels. The choroidal vessels line the outside of the eye and supply nourishment to the photoreceptors and outer retina, while the retinal vessels supply inner retinal layers with blood. Retinal capillaries, the smallest blood vessels in the eye, branch off from retinal arteries to form an intricate network throughout the whole retina with the exception of the foveal avascular zone (FAZ). The FAZ is the capillary-free region of the fovea that contains the foveal pit where the cones are most densely packed and are completely exposed to incoming light. Capillaries form a superficial layer in the nerve fiber layer, a second layer in the ganglion cell layer, and a third layer running deeper into the retina.

3. Brief theoretical background

3.1. AO calibration procedure

The AO concept requires a procedure for calculating actuator commands based on WFS signals relative to a defined set of zero points, so-called calibration. Both the DCAO demonstrator and the PoC prototype are calibrated using the same direct slope algorithm. The purpose is to construct an interaction matrix G by calculating the sensor response $s = [s_1, s_2,..., s_m]^T$ to a sequence of DM actuator commands $c = [c_1, c_2,..., c_n]^T$. Here s is a vector of measured wavefront slopes, $m/2$ is the number of subapertures, and n is the number of DM actuators. This relation is defined by

$$s = Gc, \tag{1}$$

and the interaction matrix is given by

$$G = \begin{bmatrix} \partial s_1/\partial c_1 & \partial s_1/\partial c_2 & \cdots & \partial s_1/\partial c_n \\ \partial s_2/\partial c_1 & \partial s_2/\partial c_2 & \cdots & \partial s_2/\partial c_n \\ \vdots & \vdots & & \vdots \\ \partial s_m/\partial c_1 & \partial s_m/\partial c_2 & \cdots & \partial s_m/\partial c_n \end{bmatrix}. \tag{2}$$

The relation above has to be modified to allow for multiple GSs and DMs by concatenating multiple s and c vectors. In the case of five GSs and two DMs we obtain

$$\begin{pmatrix} s_1 \\ s_2 \\ s_3 \\ s_4 \\ s_5 \end{pmatrix} = G \begin{pmatrix} c_1 \\ c_2 \end{pmatrix}, \tag{3}$$

where

$$G = \begin{bmatrix} \partial s_1/\partial c_1 & \partial s_1/\partial c_2 \\ \partial s_2/\partial c_1 & \partial s_2/\partial c_2 \\ \partial s_3/\partial c_1 & \partial s_3/\partial c_2 \\ \partial s_4/\partial c_1 & \partial s_4/\partial c_2 \\ \partial s_5/\partial c_1 & \partial s_5/\partial c_2 \end{bmatrix}. \tag{4}$$

The interaction matrix G is constructed by poking each DM actuator in sequence with a positive and a negative unit poke and calculating an average response, starting with the first actuator on DM1 and ending with the last actuator on DM2. In the case of five Hartmann patterns with 129 subapertures each and two DMs with a total of 149 actuators we obtain an interaction matrix dimension of 1290×149. The reconstructor matrix G^+ is calculated using singular value decomposition (SVD) [25] since

$$G = U \Lambda V^T,\qquad(5)$$

where U is an m×m unitary matrix, Λ is an m×n diagonal matrix with nonzero diagonal elements and all other elements equal to zero, and V^T is the transpose of V, an n×n unitary matrix. The non-zero diagonal elements λ_i of Λ are the singular values of G. The pseudoinverse of G can now be computed as

$$G^+ = V \Lambda^+ U^T,\qquad(6)$$

which is also the least squares solution to Eq. (1). The diagonal values of Λ^+ are set to λ_i^{-1}, or zero if λ_i is less than a defined threshold value. Non-zero singular values correspond to correctable modes of the system. Noise sensitivity can be reduced by removing modes with very small singular values. DM actuator commands can then be calculated by matrix multiplication:

$$\begin{pmatrix} c_1 \\ c_2 \end{pmatrix} = G^+ \begin{pmatrix} s_1 \\ s_2 \\ s_3 \\ s_4 \\ s_5 \end{pmatrix}.\qquad(7)$$

However, even the most meticulous calibration of DM and WFS interaction will not yield optimal imaging performance due to non-common path errors between the wavefront sensor and the final focal plane of the imaging channel. The reduction of these effects by proper zero point calibration is therefore crucial to achieve optimal performance of an AO system. Several methods have been proposed to improve imaging performance [26-33]. The method implemented in our system is similar to the imaging sharpening method [29, 30], but a novel figure of merit is used, and the inherent singular modes of the AO system are optimized (patent pending).

3.2. Corrected field of view

In SCAO a single GS is used to measure wavefront aberrations and a single DM is used to correct the aberrations in the pupil plane. This will result in a small corrected FOV due to field dependent aberrations in the eye. However, the corrected FOV in the eye can be increased by using several GS distributed across the FOV and two or more DMs [3, 19-21]. A larger FOV than in SCAO can actually be obtained by using several GS and a single DM in the pupil plane, analogous to ground layer AO (GLAO) in astronomy [34], but the increase in FOV size and the magnitude of correction will be less than when using multiple DMs.

A relative comparison of simulated corrected FOV for the three cases of SCAO, GLAO, and DCAO in our setup is shown in Fig. 2. The simulated FOV is approximately 7×7 degrees, with a centrally positioned GS in the SCAO simulation and five GS positioned in an 'X' formation with the four peripheral GSs displaced from the central GS by a visual angle of 3.1 deg in the GLAO and DCAO simulations.

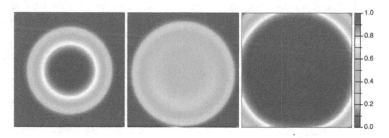

Figure 2. Zemax simulation of a corrected 7×7 deg FOV in our setup using the Liou-Brennan eye model [35] for SCAO (left), multiple GS and single DM (middle), and DCAO (right). Color bar represents simulated Strehl ratio.

4. Experimental setups

4.1. DCAO demonstrator

Only a basic description highlighting modifications to the original DCAO demonstrator will be given here. The reader is referred to [21] for a detailed description of the setup. The basic layout of the DCAO demonstrator is shown in Fig. 3.

4.1.1. DCAO demonstrator wavefront measurement and correction

Continuous, relatively broadband (to avoid speckle effects), near-infrared light (834±13 nm) from a super-luminescent diode (SLD), delivered through a 1:5 fiber splitter and five single mode fibers, is used to generate the five GS beams. The advantage of using an SLD as a source is that the short coherence length of the SLD light generates much less speckle in the Shack-Hartmann WFS spots than a coherent laser source. The end ferrules of the single mode fibers are mounted in a custom fiber holder and create an array of point sources, which are imaged via the DMs and a Badal focus corrector onto the retina. The GSs are arranged in an 'X' formation, with the four peripheral GSs displaced from the central GS by a visual angle of 3.1 deg, corresponding to a retinal separation of approximately 880 μm in an emmetropic eye.

Reflected light from the GSs passes through the optical media of the eye and emerges through the pupil as five aberrated wavefronts. After the Badal focus corrector and the two

DMs the light passes through a collimating lens array (CLA) consisting of five identical lenses, one for each GS. The five beams are focused by a lens (L7) to a common focal point (c.f. Fig. 8), collimated by a lens (L8) and individually sampled by the WFS, an arrangement consequently termed multi-reference WFS. In addition to separating the WFS Hartmann patterns as in [36] this arrangement makes it possible to filter light from all five GSs using a single pinhole (US Patent 7,639,369).

Custom written AO software for control of one or two DM and one to five GS was developed, tested, and implemented by Landell [37]. The pupil DM (DM1) will apply an identical correction for all field-points in the FOV. The second DM (DM2), positioned in a plane conjugated to a plane approximately 3 mm in front of the retina, will contribute with partially individual corrections for the five angular directions and thus compensate for non-uniform (anisoplanatic) or field-dependent aberrations. The location of DM2 was chosen to ensure an smooth correction over the FOV by allowing sufficient overlap of GS beam footprints.

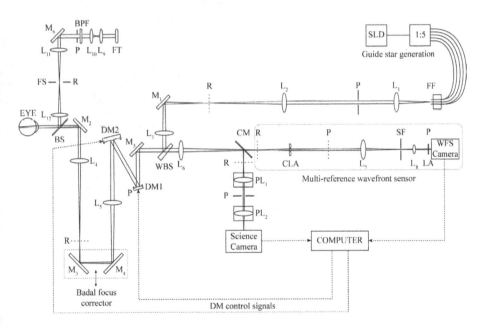

Figure 3. Basic layout of the DCAO demonstrator. Abbreviations: BPF – band-pass filter, BS – beamsplitter, CLA – collimating lens array, CM – cold mirror, DM1 – pupil DM, DM2 – field DM, FF – fiber ferrules, FS – field stop, FT – flash tube, LA – lenslet array, M – mirror, P – pupil conjugate plane, PL – photographic lens, PM – pupil mask, R – retinal conjugate plane, SF – spatial filter, SLD – superluminescent diode, WBS – wedge beamsplitter.

4.1.2. DCAO demonstrator retinal imaging

For imaging purposes, the retina is illuminated with a flash from a Xenon flash lamp, filtered by a 575±10 nm wavelength bandpass filter (BP). The narrow bandwidth of the BP is essential to minimize chromatic errors, in particular longitudinal chromatic aberration (LCA) [38] in the image plane of the retinal camera.

The illuminated field on the retina (approximately 10×10 degrees) is limited by a square field stop in a retinal conjugate plane. Visible light from the eye is reflected by a cold mirror (CM) and relayed through a pair of matched photographic lenses, chosen to minimize non-common path errors. An adjustable iris between the two photographic lenses is used to set the pupil size used for imaging, corresponding to a diameter of 6 mm at the eye.

Imaging is performed with a science grade monochromatic CCD science camera with 2048×2048 pixels and a square pixel cell size of 7.4 µm is used for imaging. The size of the CCD chip corresponds to a retinal FOV of 6.7×6.7 deg². The full width at half maximum (FWHM) of the Airy disk in the image plane at 575 nm is 15 µm and hence the image is sampled according to the Nyquist-Shannon sampling theorem (two pixels per FWHM).

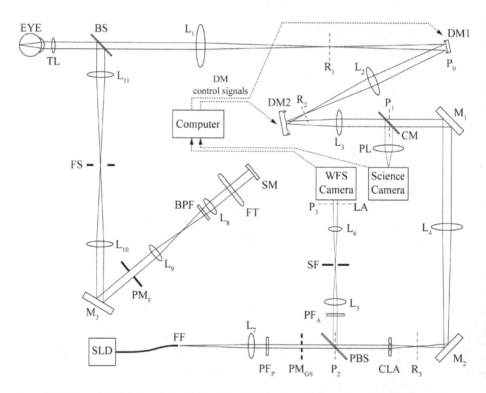

Figure 4. Basic layout of the PoC prototype. Abbreviations: BPF – band-pass filter, BS – beamsplitter, CLA – collimating lens array, CM – cold mirror, DM1 – pupil DM, DM2 – field DM, FS –field stop, FT – flash tube, LA – lenslet array, M – mirror, P – pupil conjugate plane, PBS – pellicle beamsplitter, PF_P/PF_A – polarization filters, PL – photographic lens, PM_F – flash pupil mask, PM_{GS} – GS pupil mask, R – retinal conjugate plane, SF – spatial filter, SM – spherical mirror, SLD – superluminescent diode, TL – trial lens. Fixed corrective lenses are either lens pairs or single lenses.

4.2. PoC prototype

A PoC prototype (Fig. 4) has been developed to evaluate the clinical relevance of DCAO wide-field high-resolution retinal imaging. The prototype is currently under construction and features several improvements with regards to the DCAO demonstrator. Most significant among those are that the order in which DM corrections are imposed has been changed and a novel implementation of GS creation (patent pending). The size of the PoC prototype has been greatly reduced compared with the optical table design of the DCAO demonstrator to a compact joystick operated tabletop instrument 600×170×680 mm (H×W×D) in size. The opto-mechanical layout comprises five modules: a GS generation module, a main module, a WFS module, a flash module, and an imaging module.

4.2.1. PoC GS generation module

A novel method of GS creation has been implemented in the PoC prototype, whereby the CLA that is part of the WFS is also utilized to create the GS beams. Collimated 835±10 nm SLD light from a single mode fiber is polarized (PF$_P$) and passes through a multi-aperture stop with five apertures (PM$_{GS}$) that are aligned to the five CLA lenses. Since the CLA is used for GS generation and also enables single point spatial filtering in the multi-reference WFS we have an auto-collimating arrangement that greatly reduces system complexity and alignment. The GS rays pass through standard and custom relay optics and the DMs before entering the eye, where they form five spots arranged in an 'X' formation. The four peripheral GSs are diagonally displaced from the central GS by a visual angle of 3.1 deg (880 μm) on the retina.

4.2.2. PoC main module

Residual focus and astigmatism aberrations in the DCAO demonstrator that had not been compensated for by a Badal focus corrector and trial astigmatism lenses were corrected by DM1 after passing DM2, resulting in sub-optimal DM2 performance. The PoC prototype features a correct arrangement of the DMs where reflected light from the eye, corrected by trial lenses, first passes the pupil mirror DM1 before passing the field mirror DM2.

DM1 is a Hi-Speed DM52-15 (ALPAO S.A.S., Grenoble, France), a 52 actuator magnetic DM with a 9 mm diameter optical surface and 1.5 mm actuator separation. The magnification relative to the pupil of the eye is 1.5, thus setting the effective pupil area of the instrument to 6 mm at the eye. DM2 is a Hi-Speed DM97-15 (ALPAO S.A.S., Grenoble, France), a 97 actuator magnetic DM with a 13.5 mm diameter optical surface and 1.5 mm actuator separation. GS beam footprints on DM1 and DM2 are shown in Fig. 5. The last element of the main module is a dichroic beamsplitter (CM) that reflects collimated imaging light towards the retinal camera and transmits collimated GS light towards the WFS.

As the relay optics of the main module transmits both measurement (835 nm) and imaging (575 nm) light, custom optics were designed to assure diffraction limited performance at both wavelengths (Fig. 6). Due to the ocular chromatic aberrations the bandwidth of the flash illumination bandpass filter will induce a wavelength dependent focal shift in the instrument image plane. An evaluation of the focal shift for the 575±10 nm wavelengths transmitted by the flash illumination bandpass filter using the Liou-Brennan Zemax eye model [35] yields a ±6.9 μm focal shift at the retina (Fig. 7).

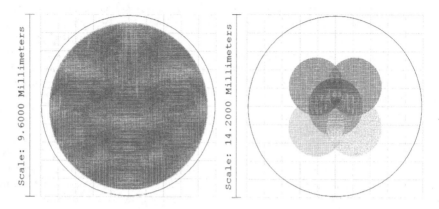

Figure 5. GS beam footprints on DM1 (left) and DM2 (right).

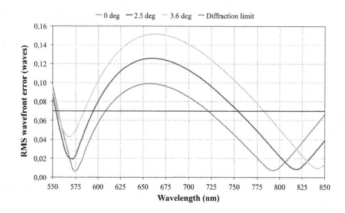

Figure 6. RMS wavefront error of the PoC main module custom relay optics at the main module exit pupil for three retinal field positions (0, 2.5, and 3.6 deg).

4.2.3. PoC WFS module

A multi-reference WFS with spatial filtering (Fig. 8) has been implemented in both the DCAO demonstrator and the PoC prototype. The design greatly reduces system complexity by implementing a single spatial filter to reduce unwanted light from parasitic source reflections and scattered light from the retina when imaging multiple Hartmann patterns with a single WFS camera.

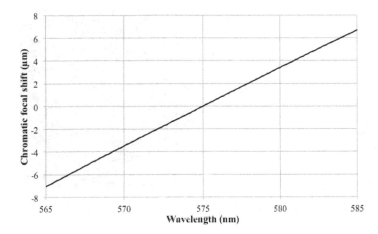

Figure 7. Chromatic focal shift over flash illumination bandpass filter bandwidth (575±10 nm) at the retina calculated using the Liou-Brennan eye model [35].

Transmitted GS light from the main module passes through the CLA and is reflected by a pellicle beam splitter. A second polarizing filter (PF_A) removes unwanted backscattered reflections from the GS generation, and a lens brings the five GS beams to a common focus where they are spatially filtered by a single aperture (SF). A collimating lens finally relays the five beams onto a lenslet array (LA) with a focal length of 3.45 mm and a lenslet pitch of 130 µm.

The monochromatic WFS CCD camera has 1388×1038 pixels with a square pixel cell size of 6.45 µm, of which a central ROI of 964×964 pixels is used for wavefront sensing. The diameter of the diffraction limited focus spot of a lenslet is 2.44 $\lambda f / d$ = 54 µm. Each spot will consequently be sampled by approximately 8×8 pixels, an oversampling that can be alleviated using pixel binning. The 6 mm pupil diameter of the eye is demagnified to 1.87 mm at the WFS and each Hartmann pattern will consequently be sampled by ~13 lenslets across the diameter (Fig. 9).

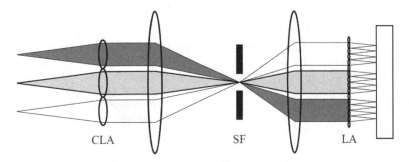

Figure 8. Schematic drawing of the multi-reference WFS with spatial filtering.

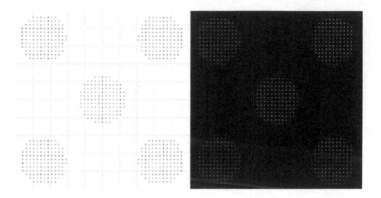

Figure 9. Zemax simulation of Hartmann spot image (left) and actual WFS image (right).

4.2.4. PoC flash and imaging modules

Retinal images are obtained by illuminating a 10×10 degree retinal field using a 4-6 ms spectral-ly filtered (575±10 nm) Xenon flash. A Canon EF 135mm f/2.0 L photographic lens is used to fo-cus reflected light from the dichroic beamsplitter onto the science camera, a 2452×2056 pixel Stingray F-504B monochromatic CCD with a square pixel cell size of 3.45 μm (Allied Vision Technologies GmbH, Stadtroda, Germany). The physical size of the full chip corresponds to a retinal FOV of 8.28×6.94 deg with a pixel resolution of 0.059 mrad (0.974 μm on the retina).

5. Retinal imaging

AO retinal imaging reveals information about retinal structures and pathology currently not available in a clinical setting. The resolution of retinal features on a cellular level offers the possibility to reveal microscopic changes during the earliest stages of a retinal disease. One of the most important future applications of this technique is consequently in clinical prac-tice where it will facilitate early diagnosis of retinal disease, follow-up of treatment effects, and follow-up of disease progression.

Both the DCAO demonstrator and the PoC prototype feature a narrow depth of focus, ap-proximately 25 μm and 9 μm in the retina, respectively. This allows for imaging of different retinal layers, from the deeper photoreceptor layer to the superficial blood vessel and nerve fiber layers. Images are flat-fielded using a low-pass filtered image to reduce uneven illumi-nation [39]. A Gaussian kernel with σ = 8 - 25 pixels is chosen depending on the imaged reti-nal layer. A smaller kernel is used for images of the photoreceptor layer and a larger kernel is consequently used for images of superficial layers. Final post-processing is performed by convolving an image with a σ = 0.75 pixel Gaussian kernel to reduce shot and readout noise. As the PoC prototype is still under construction all retinal images shown below have been acquired with the DCAO demonstrator.

5.1. Cone photoreceptor imaging

Imaging of the cone photoreceptor layer (Fig. 10) is accomplished by focusing on deeper retinal layers. The variation in cone appearance from dark to bright in Fig. 10 is an effect of the directionality [40] or waveguide nature of the cones. The retinal photoreceptor mosaic provides all information to higher visual processing stages and is many times directly or indirectly affected or disrupted by retinal disease. It is therefore of interest to study various parameters, e.g. photoreceptor spacing, density, geometry, and size, to determine the structural integrity of the mosaic. An example of this is given in Fig. 11, where the cone density of the mosaic in Fig. 10 has been calculated. Cone spacing, where possible, was obtained from power spectra of 128×128 pixel sub-regions with a 64 pixel overlap. Spacing (s) was converted to density (D) using the relation $D = \mathrm{sqrt}(3) / (2s^2)$, and the density profile was constructed by fitting a cubic spline surface to the distribution of density values.

5.2. Retinal capillary imaging

Retinal capillaries, the smallest blood vessels in the eye, are difficult to image because of their small size (down to 5 μm), low contrast, and arrangement in multiple retinal planes. Even good-quality retinal imaging fails to capture any of the finest capillary details. The preferred clinical imaging method is fluorescein angiography (FA), an invasive procedure in which a contrast agent is injected in the patient's bloodstream to enhance retinal vasculature contrast. The narrow depth of focus of both the DCAO demonstrator and the PoC prototype allows for imaging of retinal capillaries by focusing on the upper retinal layers. It is a non-invasive procedure with performance similar to FA [22]. An unfiltered camera raw image of the capillary network surrounding the fovea, the central region of the retina responsible for sharp vision, is shown in Fig. 12, and a flat-fielded image is shown in Fig. 13.

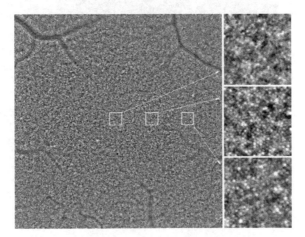

Figure 10. DCAO image of cone photoreceptor layer. Variation in cone appearance from dark to bright is an effect of the directionality or waveguide nature of cone photoreceptors.

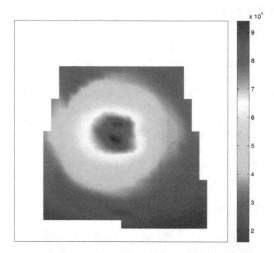

Figure 11. Cone photoreceptor density profile calculated from cone distribution in Fig. 10. Color bar represents cell density in cells/mm².

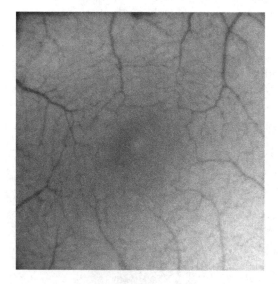

Figure 12. Camera raw DCAO image of foveal capillaries.

5.3. Nerve fiber layer imaging

Evaluation of the retinal nerve fiber layer (RNFL) is of particular interest for detecting and managing glaucoma, an eye disease that results in nerve fiber loss. Changes in the RNFL are often not detectable using red-free fundus photography until there is more than 50% nerve fiber loss [41]. Although DCAO imaging does not yet provide information about RNFL thickness it can be used to obtain images with higher resolution and contrast than red-free fundus images (Fig. 14).

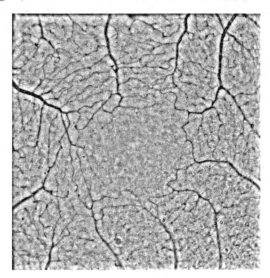

Figure 13. Image in Fig. 12 after flat-field correction. Uneven flash illumination has been reduced and retinal vessel contrast has been improved.

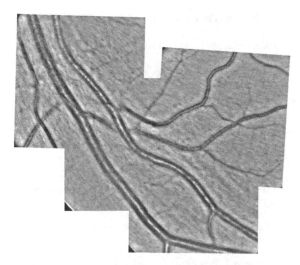

Figure 14. Montage of four DCAO images of the retinal nerve fibers and blood vessels.

6. Conclusions

In this chapter we have described the concept and practical implementation of dual-conjugate adaptive optics retinal imaging, i.e. multiconjugate adaptive optics using two deformable mirrors. Although the technique of adaptive optics is well established in the vision research community there are only a few publications on MCAO retinal imaging.

The DCAO instruments described here allow retinal features down to 2 μm to be resolved over a 7×7 degree FOV and enable tomographic imaging of retinal structures such as cone photoreceptors and retinal capillaries. We believe that this new technique has a future potential for clinical imaging at currently subclinical levels with an impact particularly important for early diagnosis of retinal diseases, follow-up of treatment effects, and follow-up of disease progression.

Acknowledgements

The authors would like to acknowledge financial support for this work from the Marcus and Amalia Wallenberg Memorial Fund (grant no. MAW 2009.0053) and from VINNOVA, the Swedish Governmental Agency for Innovation Systems (grant no. 2010-00518).

Author details

Zoran Popovic[1], Jörgen Thaung[1], Per Knutsson[1] and Mette Owner-Petersen[2]

*Address all correspondence to: zoran@oft.gu.se

1 Department of Ophthalmology, University of Gothenburg, Gothenburg, Sweden

2 Retired from the Telescope Group, Lund University, Lund, Sweden

References

[1] Babcock. HW. The Possibility of Compensating Astronomical Seeing. Publications of the Astronomical Society of the Pacific. 1953;65(386):229.

[2] Foy, Labeyrie. Feasibility of Adaptive Telescope with Laser Probe. Astronomy and Astrophysics. 1985;152(2):L29-L31.

[3] Dubinin A, Cherezova T, Belyakov A, Kudryashov A. Human Retina Imaging: Widening of High Resolution Area. Journal of Modern Optics. 2008;55(4-5):671-681.

[4] Dicke RH. Phase-Contrast Detection of Telescope Seeing Errors and Their Correction. Astrophysical Journal. 1975;198(3):605-615.

[5] Beckers JM. Increasing the Size of the Isoplanatic Patch with Multiconjugate Adaptive Optics. ESO Conference and Workshop on Very Large Telescopes and their Instrumentation; 1988; Garching, Germany: European Southern Observatory (ESO) p. 69.

[6] Beckers JM. Detailed Compensation of Atmospheric Seeing Using Multiconjugate Adaptive Optics. Roddier FJ, editor1989. 215-217 p.

[7] Ellerbroek BL. First-Order Performance Evaluation of Adaptive-Optics Systems for Atmospheric-Turbulence Compensation in Extended-Field-of-View Astronomical Telescopes. Journal of the Optical Society of America a-Optics Image Science and Vision. 1994;11(2):783-805.

[8] Fried DL, Belsher JF. Analysis of Fundamental Limits to Artificial-Guide-Star Adaptive-Optics-System Performance for Astronomical Imaging. Journal of the Optical Society of America a-Optics Image Science and Vision. 1994;11(1):277-287.

[9] Fusco T, Conan JM, Michau V, Rousset G, Mugnier LM. Isoplanatic Angle and Optimal Guide Star Separation for Multiconjugate Adaptive Optics. In: Wizinowich PL, editor. Adaptive Optical Systems Technology, Pts 1 and 22000. p. 1044-1055.

[10] Johnston DC, Welsh BM. Analysis of Multiconjugate Adaptive Optics. Journal of the Optical Society of America a-Optics Image Science and Vision. 1994;11(1):394-408.

[11] Owner-Petersen M, Goncharov A. Multiconjugate Adaptive Optics for Large Telescopes: Analytical Control of the Mirror Shapes. Journal of the Optical Society of America a-Optics Image Science and Vision. 2002;19(3):537-548.

[12] Rigaut FJ, Ellerbroek BL, Flicker R. Principles, Limitations and Performance of Multi-Conjugate Adaptive Optics. Adaptive Optical Systems Technology, Pts 1 and 2. 2000;4007:1022-1031.

[13] Berkefeld T, Soltau D, von der Luhe O. Multi-Conjugate Adaptive Optics at the Vacuum Tower Telescope, Tenerife. Adaptive Optical System Technologies Ii, Pts 1 and 2. 2003;4839:544-553.

[14] Marchetti E, Hubin N, Fedrigo E, Brynnel J, Delabre B, Donaldson R, et al. Mad the Eso Multi-Conjugate Adaptive Optics Demonstrator. Adaptive Optical System Technologies Ii, Pts 1 and 2. 2003;4839:317-328.

[15] Rimmele T, Hegwer S, Marino J, Richards K, Schmidt D, Waldmann T, et al. Solar Multi-Conjugate Adaptive Optics at the Dunn Solar Telescope. 1st Ao4elt Conference - Adaptive Optics for Extremely Large Telescopes. 2009.

[16] von der Luhe O, Berkefeld T, Soltau D. Multi-Conjugate Solar Adaptive Optics at the Vacuum Tower. Comptes Rendus Physique. 2005;6(10):1139-1147.

[17] Rigaut F, Neichel B, Boccas M, d'Orgeville C, Arriagada G, Fesquet V, et al. Gems: First on-Sky Results. Adaptive Optics Systems III; 2012: Proc. SPIE.

[18] Bedggood P, Daaboul M, Ashman R, Smith G, Metha A. Characteristics of the Human Isoplanatic Patch and Implications for Adaptive Optics Retinal Imaging. J Biomed Opt. 2008;13(2):024008. Epub 2008/05/10.

[19] Bedggood P, Metha A. System Design Considerations to Improve Isoplanatism for Adaptive Optics Retinal Imaging. Journal of the Optical Society of America a-Optics Image Science and Vision. 2010;27(11):A37-A47.

[20] Bedggood PA, Ashman R, Smith G, Metha AB. Multiconjugate Adaptive Optics Applied to an Anatomically Accurate Human Eye Model. Optics Express. 2006;14(18): 8019-8030.

[21] Thaung J, Knutsson P, Popovic Z, Owner-Petersen M. Dual-Conjugate Adaptive Optics for Wide-Field High-Resolution Retinal Imaging. Optics Express. 2009;17(6): 4454-4467.

[22] Popovic Z, Knutsson P, Thaung J, Owner-Petersen M, Sjostrand J. Noninvasive Imaging of Human Foveal Capillary Network Using Dual-Conjugate Adaptive Optics. Investigative Ophthalmology & Visual Science. 2011;52(5):2649-2655.

[23] Chan A, Duker JS, Ko TH, Fujimoto JG, Schuman JS. Normal Macular Thickness Measurements in Healthy Eyes Using Stratus Optical Coherence Tomography. Archives of Ophthalmology. 2006;124(2):193-198.

[24] Ooto S, Hangai M, Sakamoto A, Tomidokoro A, Araie M, Otani T, et al. Three-Dimensional Profile of Macular Retinal Thickness in Normal Japanese Eyes. Investigative Ophthalmology & Visual Science. 2010;51(1):465-473.

[25] Barrett HH, Myers KJ. Foundations of Image Science. Hoboken, NJ: Wiley-Interscience; 2004. xli, 1540 p. p.

[26] Blanc A, Fusco T, Hartung M, Mugnier LM, Rousset G. Calibration of Naos and Conica Static Aberrations - Application of the Phase Diversity Technique. Astronomy & Astrophysics. 2003;399(1):373-383.

[27] Carrano CJ, Olivier SS, Brase JM, Macintosh BA, An JR. Phase Retrieval Techniques for Adaptive Optics. Adaptive Optical System Technologies, Parts 1 and 2. 1998;3353:658-667.

[28] Lofdahl MG, Scharmer GB, Wei W. Calibration of a Deformable Mirror and Strehl Ratio Measurements by Use of Phase Diversity. Applied Optics. 2000;39(1):94-103.

[29] Muller RA, Buffingt.A. Real-Time Correction of Atmospherically Degraded Telescope Images through Image Sharpening. J Opt Soc Am A Opt Image Sci Vis. 1974;64(9):1200-1210.

[30] Murray L. Smart Optics: Wavefront Sensor-Less Adaptive Optics - Image Correction through Sharpness Maximisation. NUI Galway; 2006.

[31] Ren D, Rimmele TR, Hegwer S, Murray L. A Single-Mode Fiber Interferometer for the Adaptive Optics Wave-Front Test. Publications of the Astronomical Society of the Pacific. 2003;115(805):355-361.

[32] Turaga D, Holy TE. Image-Based Calibration of a Deformable Mirror in Wide-Field Microscopy. Applied Optics. 2010;49(11):2030-2040.

[33] Yoon G. Wavefront Sensing and Diagnostic Uses. In: Porter J, Queener H, Lin J, Thorn K, Awwal A, editors. Adaptive Optics for Vision Science: Principles, Practices, Design and Applications: (Wiley-Interscience; 2006. p. 63-81.

[34] Rigaut F. Ground-Conjugate Wide Field Adaptive Optics for the Elts. Beyond Conventional Adaptive Optics; 2001; Venice, Italy: European Southern Observatory, Garching p. 11-16.

[35] Liou HL, Brennan NA. Anatomically Accurate, Finite Model Eye for Optical Modeling. Journal of the Optical Society of America a-Optics Image Science and Vision. 1997;14(8):1684-1695.

[36] Goncharov AV, Dainty JC, Esposito S. Compact Multireference Wavefront Sensor Design. Opt Lett. 2005;30(20):2721-2723.

[37] Landell D. Implementation and Optimization of a Multi Conjugate Adaptive Optics Software System for Vision Research. MSc thesis. University of Gothenburg; 2005.

[38] Marcos S, Moreno E, Navarro R. The Depth-of-Field of the Human Eye from Objective and Subjective Measurements. Vision Res. 1999;39(12):2039-2049. Epub 1999/05/27.

[39] Howell SB. Handbook of Ccd Astronomy. Cambridge, U.K. ; New York: Cambridge University Press; 2000. xi, 164 p. p.

[40] Stiles WS, Crawford BH. The Luminous Efficiency of Rays Entering the Eye Pupil at Different Points. Proceedings of the Royal Society of London. 1933;112:428-450.

[41] Quigley HA, Addicks EM. Quantitative Studies of Retinal Nerve Fiber Layer Defects. Arch Ophthalmol. 1982;100(5):807-814. Epub 1982/05/01.

Devices and Techniques

Chapter 3

Devices and Techniques for Sensorless Adaptive Optics

S. Bonora, R.J. Zawadzki, G. Naletto,
U. Bortolozzo and S. Residori

Additional information is available at the end of the chapter

1. Introduction

Minimizing the aberrations is the basic concern of all the optical system designers. For this purpose, a large amount of work has been carried out and plenty of literature can be found on the subject. Until the last twenty years, the large majority of the optical design was related to "static" optical systems, where several opto-mechanical parameters, such as refractive index, shape, curvatures, etc. are slowly time dependent. In these systems, simple mechanisms can be adopted to change the relative position of one or more optical elements (for example, the secondary mirror of many astronomical telescopes), or slightly modify their shape and curvature (as in some synchrotron beamlines, where some optical surfaces are mechanically bent) to compensate defocusing. In the last years, a new type of optical systems, that we may call "dynamical", have heavily occupied the interest of optical designers, opening the possibility of working also in situations where the system environment varies rather quickly with time, either in a controlled or not-controlled way. For this class of optical systems adaptive optics (AO) with a closed loop control system has to be implemented. The correction of dynamical systems was predicted by Babcock in the 1953 [1] and, then, the first prototypes were realized in the early 70s with the purpose of satellite surveillance and launching high power laser beams trough the atmosphere [2]. The most known scientific applications of closed loop correction by means of AO is the acquisition of astronomical images in ground-based telescopes [3] and *in-vivo* imaging of cone photoreceptor mosaic by AO enhanced Fundus Cameras [4]. In astronomy, to remove the so called "seeing effect", the star light twinkling due to local dynamic variations of the atmospheric density in the air column above the telescope, it is necessary to have the real time knowledge of the wavefront of the observed object. This can be realized, for instance, by means of a Shack-Hartmann wavefront sensing device coupled to a dedicated fast algorithm which returns the mathematical description of the wavefront aberration, typically through a Zernike series decomposition [5]. Then, this information is suitably coded and passed to an AO, as a fast deformable mirror located along the optical path, that adapts its shape to compensate the time dependent aberrations. Similarly in vision science [6, 7] or retinal imaging [8-15], static and dynamic aberrations created by variation in shape of eye refractive elements and eye movements are measured by wavefront sensor, usually Schack-Hartmann and corrected by wavefront corrector, in most cases a deformable mirror. Other applications which make use of AO systems are for example: free space optical communication systems [16, 17], microscopy [18-20] or beam shaping in laser applications [21]. It is, however, rather obvious that not all AO applications have similar needs, and in particular that in some cases systems simpler than the astronomical ones can be realized. For example, in some cases there is no need to have the real time information about the aberrated wavefront: either because the aberration variation is slow [22] or because there is a specific phase which remains for a limited amount of time, as for example when correcting low order ocular aberrations "eyeglass prescription"for patient in ophthalmic diagnostics, or in optical devices in which the environmental conditions are not initially defined but the system remains stable in time [23-25]. In all these cases it can be convenient to have a simpler AO system, able to correct only the slow variations of the wavefront aberrations.

In the above mentioned cases the wavefront correction can be operated with a strong reduction in the hardware complexity, in particular by using a sensorless approach. Several techniques have been developed which use these simpler AO systems. They are generally based on the optimization of some merit function that depends on the optical system under consideration.

The algorithms for the sensorless correction can be divided into two main classes: the stochastic and the image-based ones. In the first class, the system is optimized starting from a random set and, then, applying an iterative selection of the best solutions. These algorithms have the advantage of not requiring any preliminary information about the system but they take a lot of time for converging. Many algorithms using this approach have been written and exploited successfully in different fields. Among them the most popular are: genetic algorithms [18, 26, 24], simulated annealing [13], simplex or ant colonies [27]. These approaches have the drawback of requiring a rather long computation time, or many iterations before converging, taking up to several minutes before reaching the desired system optimization.

Other sensorless techniques can be realized by analyzing some specific known feature, either intrinsic to the system or artificially introduced. An example of the latter case can be found in [28-29]. With respect to classical AO systems, the sensorless approach offers the advantage of not needing the wavefront sensor: this reduces the cost of the instrument and avoids all the problems related to maintaining the performance of such a device once installed and aligned. However, the absence of the wavefront sensor implies also some limitations, for instance, a much longer time before reaching an optimal image quality, or a final image not perfectly optimized. Clearly, the required final result and the available resources are the key elements driving the choice towards one system or another. In section 2, we will explain in detail the genetic algorithm and the ant colonies optimization process, while providing a few examples of their application in optical experimental setups.

The image-based algorithms will be explained in section 3, together with a few examples of recently reported successful applications in optical experiments. New devices useful to generate the bias aberrations will also be presented.

2. Stochastic algorithms for sensorless correction

2.1. Genetic algorithm

A genetic algorithm [30] searches the solution of a problem by simulating the evolution process. Starting from a population of possible solutions, it saves some of the strongest elements, that are the only ones selected to survive, and, thus, are able to reproduce themselves giving rise to the next generations. In general, the inferior individuals can survive and reproduce with a smaller probability.

This strategy allows solving a large class of problems without any initial hypothesis or preliminary knowledge. Its effectiveness was demonstrated in many experimental setups, as will be discussed in the following paragraphs.

The main steps of a genetic algorithm are depicted in Table 1 and in Fig. 1.

Starting random Population
1_selection function
2_reproduction function
3_evaluate population
4_repeat from step 1

Table 1. Main steps required by a genetic algorithm.

The initial population is chosen randomly in the whole set of possible solutions. The selection function can be either probabilistic or deterministic. In the probabilistic case, the strongest elements have more chances of being selected and of reproducing to the next generation. This decreases the possibility of falling in a "local" maximum solution.

The reproduction function creates new individuals from the old population. There are two kinds of functions: crossover and mutations.

CrossOver functions: they mix the genes of the two parents by slightly modifying them and by obtaining two sons.

Example: EuristicXOver:

From the parents $V_a^{(k-1)}$ and $V_b^{(k-1)}$, the children $V_a^{(k)}$ and $V_b^{(k)}$ are generated by the following rule:

$$V_a^{(k)} = V_a^{(k-1)} + r\left(V_b^{(k-1)} - V_a^{(k-1)}\right)$$

$$V_b^{(k)} = V_b^{(k-1)}$$

Mutations functions: the genes of the parent are randomly modified.

Example: Uniform Mutation:

The mutation take an element $V_{cj}^{(k-1)}$ and mutate it in a new one by the rule:

$$V_{cj}^{(k)} = \begin{cases} V_{cj}^{(k-1)} + w(k)\left(1 - V_{cj}^{(k-1)}\right) & \text{if } rand > 0.5 \\ V_{cj}^{(k-1)} + w(k)V_{cj}^{(k-1)} & \text{if } rand < 0.5 \end{cases}$$

where w(k) is weight function which decreases with the iteration k.

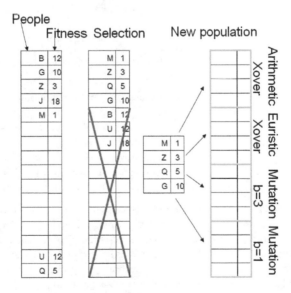

Figure 1. Diagram representing the genetic algorithm principle. The algorithm starts from a random population and then each individual is measured and the population is sorted according to its fitness. Then, some of the best individuals are selected for the generation of the next population.

2.1.1. Application example: Laser focalization

The intensity of a laser in its focal spot is largely dependent on the quality of the focal point, and this effect is even stronger in nonlinear optics. Often, in laser systems it is not simple to reach an optimal alignment, so that AO devices can be very useful in these cases.

For example, in ref. [24] it was demonstrated how an AO sensorless optimization based on a genetic algorithm can largely enhance the XUV high-order harmonics (HH) generated by the interaction of an ultrafast laser and a gas jet.

The AO system was composed by an electrostatic deformable mirror (Okotech) placed before the interaction chamber as illustrated in Fig. 2. The feedback for the genetic algorithm was the photon flux at the shortest wavelengths acquired placing a photomultiplier tube at the XUV spectrometer output.

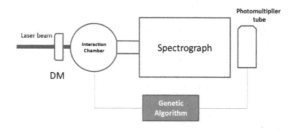

Figure 2. Experimental setup for the optimization of a laser focalization used for high order harmonics generation in ultrafast nonlinear optics. The pulsed laser beam interacts with a gas jet in the interaction chamber. The photomultiplier tube collects the signal from the spectrograph and feeds the genetic algorithm that drives the deformable mirror DM.

The laser pulse was generated by a Ti:S CPA laser system with a hollow-fiber to realize the compression of the pulse duration. The typical values used in the experiment are 6 fs of duration, 200 μJ of pulse energy, at 1 kHz repetition rate (all the experimental details are described in Villoresi et al. 2004). The focusing of the laser pulses on the gas jet, after the modifications introduced by the Deformable Mirror (DM), is obtained by means of a 250 mm focal length spherical mirror. The spectrometer that analyzes the HHs beam is based on a flat varied-line-spacing grazing-incidence grating with two toroidal mirrors.

The real-time acquisition of the spectral intensity is realized by the combination of a solar-blind open microchannel-plate (MCP) with MgF_2 photocathode and a phosphor screen placed on the spectrometer focal plane, which converts the HHs XUV spectrum in the visible, and by a photomultiplier which acquires a HHs spectral interval selected with a slit. In this way, the single-shot intensity of a single harmonic, or group of harmonics, is used as feedback by the algorithm. A separate optical channel acquires in parallel the image of all at the MCP, from which the HHs spectrum is obtained.

The genetic algorithm used a population of 80 individuals, with a deterministic selection rule that saved the 13 best ones. Both mutations and crossover were used. The results showed an increase of the XUV photons by a factor of 5 when the algorithm was applied. Moreover, the cutoff region moved to shorter wavelengths as reported in Fig 3. The optimization process took about 20 iterations to converge.

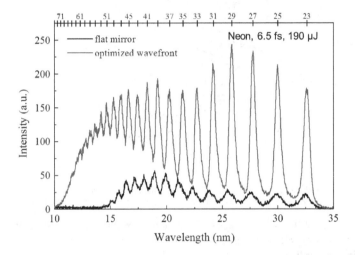

Figure 3. Result of the experimental optimization of the high order harmonics generation spectra in the case of the flat AO mirror (black line) and in the case of the optimized wavefront (red line).

2.2. Ant colonies

Ant colonies, in natural world, search the food by walking randomly. After having found it, they return to their colony leaving down a pheromone trail. If other ants cross the same trail they will not walk randomly but they will likely follow it and will reinforce the pheromone trail. The more ants will find food at the end of the trail, the more pheromone will mark it. However, since the pheromone evaporates reducing its strength, the described process will make the shortest path which will be the one with the highest density of pheromone, so providing a selection among all the possible paths, as illustrated in Fig. 4.

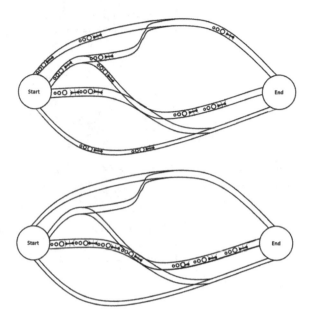

Figure 4. Ants start randomly their search for food, then the shortest path gets the higher content of pheromone. Finally, the ants will follow with larger probability the path having the highest content of pheromone.

The main essence of the Ant Colonies optimization algorithm [27] is to simulate the ant behavior for the optimization of a given problem. The algorithm steps necessary for running the optimization are listed in Table 2.

1_Set the initial ants position on the trail
2_Compute the paths length
3_Update pheromone
4_Move the ants
5_Go to step 2

Table 2. Steps of an ant colony algorithm.

As an example we show in Fig. 5 the simulation of the application of the ant colony strategy to a deformable mirror with 32 actuators and 8 bits control. In this example the actuators and their control values are the domain in which the ants can move. In the simulation the shortest path is a parabolic function, which is represented by the red line. Fig. 5 (top) shows the initial random pheromone distribution, while Fig. 5 (bottom) shows the pheromone distribution at the end of the optimization process.

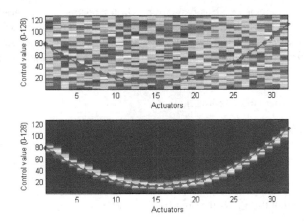

Figure 5. Implementation of an ant colony strategy for the optimization of a deformable mirror with 32 actuators and 8 bit control. The red curve represents the shortest (optimized) path. The top panel shows the initial random pheromone distribution while the bottom panel shows the pheromone at the end of the selection process.

2.2.1. Application example: Quantum optics

The quality of an optical wavefront plays an important role in Spontaneous Down Conversion (SPDC) process. As demonstrated by [31] the use of a deformable mirror can enhance the generation of photon pairs acting on the wavefront before the generation takes place in the nonlinear crystal. In that system the optimization was carried out by the use of an electrostatic DM (PAN, Adaptica srl) and the application of the ant colonies algorithm.

In the experiment, the pump beam is reflected by the DM to a BBO type-I nonlinear crystal. Then, the degenerate SPDC photons at 808 nm are selected and measured by a high efficiency SPADs (Single Photon Avalanche Diode). Since the wavefront has a strong effect on the downconverted light, it can strongly affect the coupling in the fibers of the SPAD detectors. The feedback for the algorithm imposed the condition of photon coincidences. It was demonstrated in the experiment that the coincidences rate was increased by about 20% when the optimization algorithm was applied. The algorithm used about 80 ants and the convergence took place in about 800 iterations.

3. Image based algorithms

Although the stochastic optimization algorithms have been demonstrated to represent important tools for optical experiments, new techniques, which demonstrated to be more effective, have recently been introduced. The use of a modal approach, based on the application of bias aberrations and of a suitable metrics, sorted out some of the limitations of the search algorithms, such as the long convergence time and the need of a training for the determination of the algorithm parameters. This new approach demonstrated to be effective both in visual optics and in laser optimization, as described later in this section. The arbitrary generation of aberrations can be achieved through the use of deformable mirrors, either thanks to a preliminary calibration of them or through the design of a suitable new class of wavefront correctors [32].

3.1. Devices for sensorless modal correction

Electrostatic membrane deformable mirrors rely on the electrostatic pressure between an actuator pad array and a thin metalized membrane [33]. Thus, the more the actuators the better the wavefront resolution that the mirror can control. The use of these deformable mirrors is, then, subjected to the acquisition of the deformation generated by each electrode. On the other hand, this kind of DMs can also be used with the optimization algorithms. The drawback, in this case, is that the higher the number of actuators the longer will take to the algorithm to converge.

Recently, a new type of deformable mirrors suitable for the direct generation of aberrated wavefronts was designed. The modal membrane deformable mirror, MDM, relies on the use of a graphite layer electrode arrangement (see Fig. 6) for the generation of a continuous distribution of the electric field which allows the generation of the low order aberrations (defocus, astigmatism, coma) and of the spherical aberration.

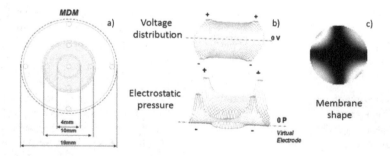

Figure 6. Electrostatic modal membrane deformable mirror, MDM. (a) Layout of the electrodes of the MDM; (b) voltage and electrostatic pressure distribution which generates the astigmatism shape illustrated in the interferogram shown in (c).

The MDM has already been demonstrated to be effective in several fields, as laser focalization [32], image sharpening and Optical Coherence Tomography (OCT), as it will be discussed later.

Another device for the generation of aberrations is the PhotoControlled Deformable Mirror (PCDM), which is schematically represented in Fig. 7. This deformable mirror [35, 36] is composed of an electrostatic membrane while the actuator pad array is replaced by a photoconductive material. Thus, the membrane shape depends on the light pattern projected on the photoconductor. Arbitrary actuator pads can be conveniently achieved by illuminating the photoconductive side of the mirror with a commercially available Digital Light Processing (DLP) hand-held projector.

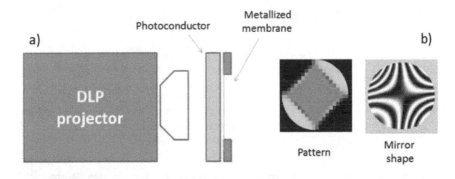

Figure 7. Photo-controlled deformable membrane mirror, PCDM. a) Schematic representation of the PCDM and the projection system allowing to achieve arbitrary actuator pads. b) Left: layout of the electrode pattern; right: correspondingly generated mirror shape; as an example the electrode pattern was chosen to generate astigmatism.

The calculation of the electrode pattern that generates a determined aberration is composed of the following steps:

a. division of the projector area into small subsets (i.e. 40 × 40);

b. calculation of the membrane shape for each of the 40 × 40 pixels, solving the Poisson equation by the iterative methods;

c. determination of the pattern by pseudoinversion of the matrix determined at point b.

A few examples of the realized electrode patterns are shown in Fig. 8, together with the corresponding measurements of the aberrated wavefronts.

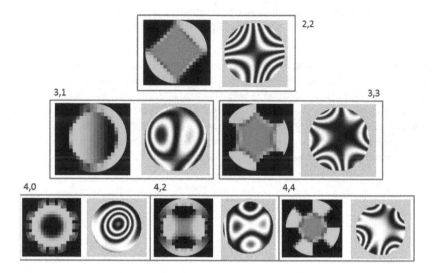

Figure 8. Generation of the first four Zernike orders with the photocontrolled deformable mirror; the light patterns necessary for their generation are on the left; the obtained corresponding interferograms are on the right.

We proved that the image quality can be considerably improved by using these adaptive devices in an image sharpening setup. For example, the MDM allowed achieving a significant image sharpening with just about 35 measurements, as illustrated in Fig. 9.

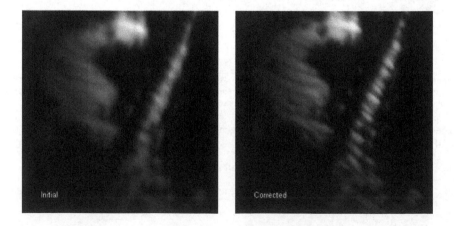

Figure 9. Optimization of an image deteriorated by aberrations; left: initial image; right: image corrected by the MDM (after 35 measurements).

3.2. Optimization of low spatial frequencies

The sharpness of an optical imaging system depends strongly on the wavefront quality. Recently [28] demonstrated that the low spatial frequency in an image can be used as a metric to perform the optimization. The process takes place by the acquisition of a series of images with the application of a predetermined aberration. The images, then, contain the information about the corrections which have to be applied to cancel the aberrations. This technique is very powerful, especially if coupled to the Lukosz modes aberration expansion. The Lukosz modes are similar to the Zernike polynomials: the difference is that the Zernike polynomials are normalized such that a coefficient of value 1 generates a wavefront with a variance of 1 rad², while the Lukosz functions are normalized such that a value 1 coefficient corresponds to a rms spot radius of $\lambda/(2\pi NA)$, where λ is the wavelength and NA is the numerical aperture of the focusing lens.

The peculiarity of this expansion is that the effect of the Lukosz polynomials coefficients $\{a_i\}$, on the image sharpness $I(a_i)$ is quadratic:

$$I(a_i) \approx \sum_i a_i^2.$$

This implies that the optimization of each mode can be performed independently and requires just the acquisition of three images. Then, the best point for each aberration can be found by interpolating the result with a quadratic function.

3.3. Point Spread Function (PSF) optimization

Another example of application of wavefront sensorless AO [29] consists in projecting a known point-like source through the optical system under test and then analyzing its image by means of a suitable software [36]. With the information obtained by the analysis of the point source image, the shape of a deformable mirror inserted along the optical path is modified. This process is iteratively repeated through a defined hierarchy, to gradually remove the optical aberrations.

With this technique, the point source image to be analyzed is not directly available and has to be somehow created. As an example, in the case of a fundus camera dedicated to the observation of the human retina, an illuminated pinhole can be projected on the retina itself through a dedicated optical path; this is a standard technique for this type of applications and is not going to introduce a significant complexity in the system. The light that is back-diffused by the retinal fundus acts as a point source, and its wavefront can be analyzed to estimate the aberrations present along the optical path from the retina to the detector.

3.3.1. Application of the PSF optimization in a visual optics setup

The closed loop method of correcting the aberrations of an optical system has been verified to be very stable, at least with respect to possible misalignments of the deformable mirror or aging of the mirror membrane that has been used. This stability is inherent in the adopted approach to the problem, which is less ambitious than correcting the wavefront aberration.

The described technique has been verified by means of the rather simple optical setup shown in Fig. 10. The radiation emitted by a LED diode source (SOU) is condensed by a microscope objective lens (L_{cond}) on a pinhole (PH). The radiation emerging from the pinhole is collected by a zoom collimating lens (L_{coll}). The collimated beam passes through a diaphragm (DIA) and a beam splitter (BS) and impinges normally onto a deformable mirror (M_{def}). After reflections from M_{def} and BS, the beam is compressed by an a-focal Newtonian system (L_{comp}^1 and L_{comp}^2) and can, then, follow two different paths: either a) a focusing two-lens system L_{foc} that makes the image of the pinhole on a CMOS digital camera (DET), or b) a flip mirror (M_{flip}) which deviates the beam on a wavefront analyzer (WFA). The latter has been used to measure the wavefront aberrations before and after the correction performed by the DM. With this system, both by varying the focal length of L_{coll} and tilting L_{comp}^1, it was possible to introduce controlled amounts of aberrations on the nominal pinhole image. Then, by the suitable image analysis and consequent estimate of the aberrations, the parameters needed to drive the deformable mirror to improve the image quality have been derived.

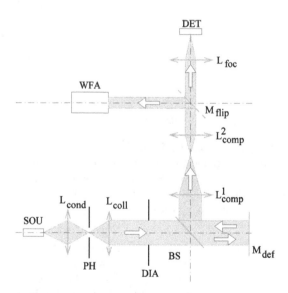

Figure 10. Schematic representation of the optical setup used for testing the capability of correcting the system aberrations with a sensorless technique. SOU: source LED diode; M_{def}: deformable mirror; DET: CMOS camera for detection; WFA: wavefront analyzer. See text for a complete description.

Even if the apparatus performance was constrained by the limited unidirectional sag of the deformable mirror, the obtained results proof the principle of the adopted methodology. This is clearly demonstrated in Fig. 11, which shows the wavefront error measured with the WFA before and after the deformable mirror correction for three different cases. From these graphs, and more quantitatively from the detailed analysis described in [29], it can be seen that a RMS wavefront error as low as $\lambda/10$ (@527.5 nm) can be obtained, which is a significant result for a sensorless AO system. The correction was not particularly effective only in those cases in which the unidirectionality of the mirror deformation did not allow aberration compensation, as in case of astigmatism. However, with a different choice of AO system, the system is very effective in identifying and correcting aberrations.

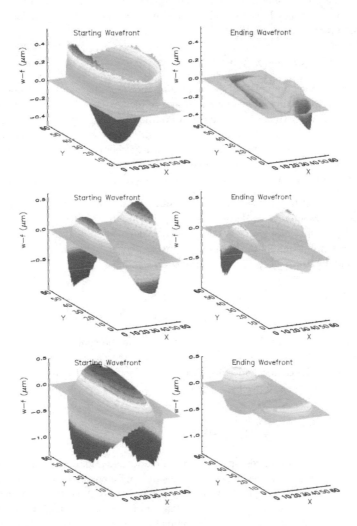

Figure 11. Wavefront plots (obtained with the wavefront analysis, WFA) before and after the correction applied by the deformable mirror for three considered cases. Top: the main aberration is defocus; Middle: the main aberrations are astigmatism and coma; Bottom: the main aberrations are defocus and astigmatism. The blue dashed lines over plotted to the Z axis represent the total wavefront excursion.

It has to be mentioned that these tests verified that the image analysis algorithm takes less than 100 cycles to reach the optimal condition; since one cycles takes approximately 1/20 - 1/25 s on a standard computer, the whole optimization takes just 4-5 s. Comparing this time with the typical times necessary to optimize other sensorless AO systems, it is evident the significant advantage of this technique, once implemented. The only limitation of this technique is that starting PSF image should give enough signal. In fact, if the point source image is too spread out, the signal to noise ratio can be very poor, substantially inhibiting the system to make a correct image analysis.

3.4. Optical Coherence Tomography (OCT)

Optical coherence tomography, OCT, is an imaging modality allowing acquisition of micrometer-resolution three-dimensional images from the inside of optical scattering media (e.g. biological tissue). OCT is analogous to ultrasound imaging, except that it makes use of light instead of sound. It relies on detecting interferometric signal created by the light back scattered from the sample and from a reference arm in a Michelson or Mach-Zehnder interferometer. OCT has many applications in biology and medicine and can be treated as a sort of optical biopsy without requirement of tissue processing for microscopic examination.

One of the interesting features of OCT is that, unlike in most optical imaging techniques, the axial and lateral resolutions are decoupled, thus allowing for an improved axial resolution, which is independent of transverse resolution. The axial resolution Δz is determined by the roundtrip coherence length of the light source and can be calculated from the central wavelength (λ_0) and the bandwidth ($\Delta\lambda$) of the light source as [37]:

$$\Delta z = \frac{2\ln 2}{\pi} \frac{\lambda_0^2}{\Delta\lambda}.$$

The lateral resolution (Δx) in OCT is defined similarly to the confocal scanning laser ophthalmoscopy (cSLO), since OCT is based on a confocal imaging scheme. In many imaging systems, however, the confocal aperture exceeds the size of the Airy disc, which degrades the resolution to the value known from microscopy, i.e. [38]:

$$\Delta x = 1.22\lambda \frac{f}{D}.$$

Therefore, as for standard microscopy, AO enhanced devices might be necessary to achieve diffraction limited transverse resolution. As a result, only a combination of OCT with AO has the potential to achieve high and isotropic volumetric resolution. The use of broadband light sources that are necessary for OCT and the complexity of both the AO and the OCT technique, make the combination very challenging [39]. In general, any AO-OCT instrument can be divided into two subsystems: an adaptive optics subsystem, with wavefront sensing and wavefront correction, and an interferometric OCT subsystem. In every implementation of AO-OCT all the elements of the AO subsystem are located in the sample arm of the OCT interferometer. Indeed, there is no need to have AO correction in the reference arm because aberrations introduced within this part of the system will not influence the transverse resolution of the image. In most of the AO-OCT systems, a Shack–Hartmann wavefront sensor is used to measure aberrations and, then, to control adaptive optics correction.

Bonora and Zawadzki recently demonstrated that sensorless correction can be implemented in optical coherence tomography by using a specially developed resistive deformable mirror. This novel modal deformable mirror, MDM, was successfully employed in the UC Davis AO-OCT system to image static samples, test targets and tissue phantoms. Fig. 12 shows a schematic representation of the sensorless AO-OCT system used in the experiments.

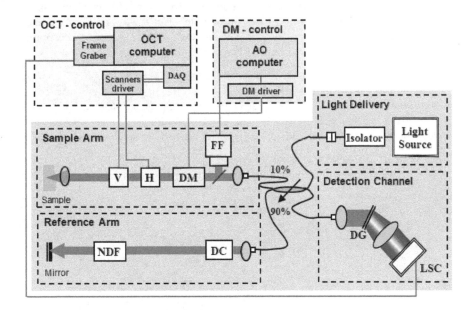

Figure 12. Schematic representation of the system for sensorless adaptive optics - optical coherence tomography. Note that there is no wavefront sensor in the sample arm. The far-field camera (FF) is used to check if the AO correction generates improved focal spots. DM : deformable mirror; V: vertical mirror galvanometer; H: horizontal mirror galvanometer. In the reference arm: NDF is a neutral density filter. The detection channel comprises a grating (DG) and a linear CCD detector (LSC). The quality of the image acquired with the OCT detection channel is used to search for DM shapes that correct aberrations in the imaged sample. The imaging system used to acquire the data was developed in the Vision Science and Advanced Retinal Imaging Laboratory (VSRI). Details of the OCT system components can be found in [40]. Here, we briefly describe the main characteristics of the system. In the current configuration, the light source for OCT was a superluminescent diode (Broadlighter) operating at 836 nm and with a 112 nm spectral bandwidth (Superlum LTD), allowing to achieve a 3.5 μm axial resolution. The beam diameter at the last imaging objective was 6.7 mm, allowing for up to 10 μm lateral resolution when a 50 mm focal length imaging objective was used. The AO correction was optimized by using the intensity of the AO-OCT en-face projection views during the volumetric data acquisition. In the current system configuration, we have used about 9 mm diameter of the modal deformable mirror. The light reflected from the sample is combined with the light from the reference mirror, and then sent to a spectrometer. There, a CCD line detector acquires the OCT spectrum.

To test the performance of our sensorless AO-OCT system, we evaluated the image quality of a sample, consisting of a USAF resolution test chart with an adhesive tape glued to its front side, after insertion of a trial lens with 0.5 Diopter astigmatism in front of the imaging objective. We were able to achieve improved resolution by using the following merit function S [41] on the OCT en-face projection images

$$S = \int I^2(x, y) dx dy,$$

where I(x,y) is the intensity in the OCT en-face image plane. This approach is simillar to PSF optimization. In fiber based OCT systems single mode fiber introduces OCT beam to the sample and also act as detector for back scattered light. Therefore we have a point source that is imaged by the optical system and the confocal pinhole that allows direct mesurment of light intensity trougput by the system. As expected, the algorithm performed the optimization by adjusting only defocus and astigmatism (see Fig. 13).

Fig. 14 shows some examples of the en-face projection views extracted from OCT volumes: there are the initial view acquired from the sample, and three improved views after correction of additional aberrations, namely, defocus and two astigmatisms. Clearly, at each correction step the images of the test target get sharper. Additionally, the features of the adhesive tape attached to the back of the Air Force test target become more visible as well.

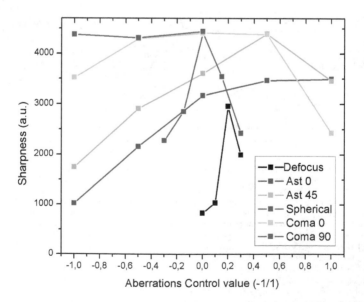

Figure 13. Graph of the Merit Function of AO-OCT images for different values of aberrations generated by the modal deformable mirror. Note that higher values correspond to better AO-corrections.

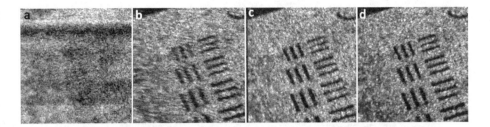

Figure 14. En-face projection views of the AO-OCT images of the test target for the best corrected values of the Zernike coefficients; (a) before correction, (b) after defocus correction, (c) after defocus and Ast 0° correction, (d) after defocus, Ast 0° and Ast 45° correction.

These recent results demonstrate that wavefront sensorless control is a viable option for imaging biological structures for which AO cannot establish a reliable wavefront that could be corrected by a wavefront corrector. Future refinements of this technique, beyond the simple implementation presented in this chapter, should allow its extension to in-vivo applications. An example of sensorless adaptive optics scanning laser ophthalmoscopy (AO-SLO) for imaging in-vivo human retina has been recently presented [42].

3.5. Laser process optimization

Similarly to the optimization process presented in section 2.1.1 [24], we report here about the optimization of a laser process by the use of a sensorless AO [43]. In the former case, the generation of harmonics from an ultrafast laser was improved by the use of a genetic algorithm. In the latter case, an algorithm derived from the image-based procedure was employed in conjunction with the use of a MDM deformable mirror similar to the one described in section 3.1. The advantages in terms of experimental complexity and convergence time are discussed in the given reference.

In the sensorless case, the laser source was a tunable high energy mid-IR (1.2μm-1.6μm) optical parametric amplifier with 10 Hz repetition rate [44]. The harmonics of the laser were generated by the interaction of the laser pulses with a krypton gas jet. In this system, the infrared pulses and the slow repetition rate made inconvenient, respectively, the use of a wavefront sensor and of an optimization algorithm needing hundreds of iterations.

The experimental setup used for this application is illustrated in Fig. 15. To demonstrate the easiness of integrating the sensorless AO device within the experiment, the optical path before the DM is shown with a dotted line. The additional elements are simply a plane mirror and a resistive MDM, which have been introduced without any complex operations. The system optimization consisted in the increase of the harmonic signal detected by the photomultiplier at the output of the monochromator. The obtained result is illustrated in Fig. 16, where it is possible to see that the photon flux on the photomultiplier is doubled with respect to the one obtained after the correction of the defocus.

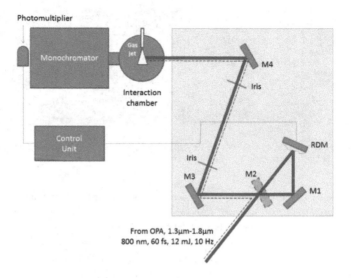

Figure 15. Experimental setup for the generation of harmonics from a femtosecond tunable high-energy mid-IR optical parametric amplifier, OPA. Dotted line: optical path before the insertion of the MDM. Red line: optical path realized for the experiment with the deformable mirror.

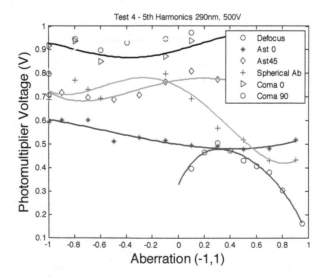

Figure 16. Optimization of the voltage generated by the photomultiplier over a 50 Ω load for the 5th harmonic at 290 nm, obtained by the use of krypton gas.

4. Conclusions

In adaptive optics the choice of the optimal correction strategy depends on the required application, desired image quality, and affordable complexity/cost of the final system. In this context, sensorless adaptive optics provides several solutions, most of them implementable at a simplified and relatively low-cost level, that can be exploited for a wide range of applications.

We have presented here both a review of the most diffused systems used in sensorless adaptive optics and some recently developed algorithms and devices. Essentially, two different approaches are employed: those based on random search and the subsequent application of evolutionary strategies, and those based on the application of some bias aberration. In general, the second class of algorithms present a faster convergence.

We have shown several application examples in different fields, such as the optimization of ultrafast nonlinear optical systems for the generation of high order harmonics, the image sharpening in microscopy applications and the enhancement of optical coherent tomography.

Sensorless adaptive optics appears, therefore, as having a great potential for finding new applications in current and future technologies. The continuous improvement of the optimization algorithms and development of novel deformable mirror devices, make the integration of AO into various optical systems increasingly easier. Particularly, the conjunction of sensorless AO with OCT might open the way to a new generation of diagnostic imaging.

Author details

S. Bonora[1], R.J. Zawadzki[2], G. Naletto[1,3], U. Bortolozzo[4] and S. Residori[4]

1 CNR-IFN, Laboratory for UV and X-Ray and Optical Research, Padova, Italy

2 VSRI, Department of Ophthalmology and Vision Science, University of California Davis, Sacramento, CA, USA

3 Department of Information Engineering, University of Padova, Padova, Italy

4 INLN, Université de Nice-Sophia Antipolis, CNRS, France

References

[1] Babcock H.W., The Possibility of Compensating Astronomical Seeing, Publication of the Astronomical Society of the Pacific Vol.65, No. 386, pp. 229, (1953).

[2] Tyson R., Tharp J., Canning D., Measurement of the bit-error rate of an adaptive optics, free-space laser communications system, part 2: multichannel configuration, aberration characterization, and closed-loop results, Optical Engineering Vol. 44, No. 9, pp. 096,003-1 096,003-6, (2005).

[3] Hardy J.W., Adaptive Optics for Astronomical Telescopes, (Oxford University Press, ISBN-10: 0195090195, USA, 1998).

[4] Liang J., Williams D., Miller D., Supernormal vision and high resolution retinal imaging through adaptive optics, Journal of the Optical Society of America A, Vol. 14, No. 11, pp. 2884-2892, (1997).

[5] Irwan R., Lane R., Analysis of optimal centroid estimation applied to Shack Hartmann sensing, Applied Optics Vol. 38, No. 32, pp. 6737-6743, (1999).

[6] Porter J. , Queener H. , Lin J., Thorn K. E., Awwal A., Adaptive Optics for Vision Science: Principles, Practices, Design and Applications (Wiley, 2006).

[7] Roorda A., Adaptive optics for studying visual function: A comprehensive review, Journal of Vision, Vol. 11, No. 5, pp. 1-21 (2011).

[8] Liang J., Williams D., Aberrations and retinal image quality of the normal human eye, Journal of the Optical Society of America A, Vol. 14, No. 11, pp. 2873-2883, (1997).

[9] Zhu L., Sun P., Bartsch D., Freeman W., Fainman Y., Adaptive control of a micro-machined continuous-membrane deformable mirror for aberration compensation, Applied Optics, Vol. 38, No. 1, pp. 168-176, (1999).

[10] Le Gargasson J.-F., Glanc M., Léna P., Retinal imaging with adaptive optics, ComptesRendus de l'Acadèmie des Sciences - Series IV - Physics 2, pp. 1131-1138, (2001).

[11] Roorda A., Romero-Borja F., Donnelly W., Queener H., Hebert T., Campbell M., Adaptive optics scanning laser ophthalmoscopy, Optics Express, Vol. 10, No. 9, pp. 405-412, (2002).

[12] Zawadzki R., Jones S., Olivier S., Zhao M., Bower B., Izatt J., Choi S., Laut S., Werner J, Adaptive-optics optical coherence tomography for high-resolution and high-speed 3D retinal in vivo imaging, Optics Express, Vol. 13, No. 21, pp. 8532-8546, (2005).

[13] Zommer S., Ribak E., Lipson S., Adler J., Simulated annealing in ocular adaptive optics, Optics Letters, Vol. 31, No. 7, pp. 1-3, (2006).

[14] Gray D., Merigan W., Wolfing J., Gee B., Porter J., Dubra A., Twietmeyer T., Ahamd K., Tumbar R., Reinholz F., Williams D., In vivo fluorescence imaging of primate retinal ganglion cells and retinal pigment epithelial cells, Optics Express, Vol. 14, No. 16, pp. 7144-7158, (2006).

[15] Fernandez E., Vabre L., Adaptive optics with a magnetic deformable mirror: application in the human eye, Optics Express, Vol. 14, .No. 20, pp. 8900-8917, (2006).

[16] Tyson R., 1999, Adaptive Optics Engineering Handbook , CRC Press, ISBN-10: 0824782755, New York USA

[17] Tyson R., Tharp J., Canning D., Measurement of the bit-error rate of an adaptive optics, free-space laser communications system, part 1: tip-tilt configuration, diagnostics, and closed-loop results, Optical Engineering Vol. 44, No. 9, pp. 096,002-1 096,002-6, (2005).

[18] Albert O., Sherman L., Mourou G., and Norris T., Smart microscope: an adaptive optics learning system for aberration correction in multiphoton confocal microscopy, Optics Letters Vol. 25, No. 1, pp. 52-54, (2000).

[19] Neil MAA., Juskaitis R., Booth M.J., Wilson T., Tanaka T., Kawata S., Adaptive Aberation correction in two-photon microscope, Journal of microscopy, Vol 200, pp 1-5-108 (2000).

[20] Booth M.J., Neil M.A.A., Juskaitis R., Wilson T., Adaptive aberration correction in a confocal microscope, PNAS, Vol. 99, No. 9. Pp. 5788-5792 (2002).

[21] Brida D., Manzoni C., Cirmi G., Marangoni M.,, Bonora S., Villoresi P., De Silvestri S., Cerullo G., (2010), Few-optical-cycle pulses tunable from the visible to the mid-infrared by optical parametric amplifiers, Journal of Optics, Vol. 12, No. 1, (January 2010), 2040-8978

[22] Okada T., Ebata K., Shiozaki M., Kyotani T., Tsuboi A., Sawada M., Fukushima M., Development of adaptive mirror for CO_2 laser, in High-Power Lasers in Manufacturing, X. Chen, T. Fujioka, and A. Matsunawa, eds., Vol. 3888 of SPIE Proc., pp. 509-520, (2000).

[23] Jackel S., Moshe I., Adaptive compensation of lower order thermal aberrations in concave-convex power oscillators under variable pump conditions, Optical Engineering Vol. 39, No. 09, pp. 2330-2337, (2000).

[24] Villoresi P., Bonora S., Pascolini M., Poletto L., Tondello G., Vozzi C., Nisoli M., Sansone G., Stagira S., De Silvestri S., Optimization of high-order-harmonic generation by adaptive control of sub-10 fs pulse wavefront, Optics Letters, Vol. 29, No.2, pp. 0146-9592, (2004).

[25] Zacharias R., Beer N., Bliss E., Burkhart S., Cohen S., Sutton S., Atta R.V., Winters S., Salmon J.T., Stolz M. L. C., Pigg D., Arnold T., Alignment and wavefront control systems of the National Ignition Facility, Optical Engineering, Vol. 43, No. 12, pp. 2873-2884, (2004).

[26] Gonté F., Courteville A., Dandliker R., Optimization of single-mode fiber coupling efficiency with an adaptive membrane mirror, Optical Engineering, Vol. 41, No. 5, pp. 1073-1076, (2002).

[27] Bonabeau E., Dorigo M. and Theraulaz G., Inspiration for optimization from social insect behaviour, Nature Vol. 46, pp. 39-42, (2000).

[28] Debarre D., Booth M.J. and Wilson T., Image based adaptive optics through optimisation of low spatial frequencies, Optics Express, Vol. 15, No. 13, pp. 8176-8190, (2007).

[29] Naletto G., Frassetto F., Codogno N., Grisan E., Bonora S., Da Deppo V., Ruggeri A. (2007), No wavefront sensor adaptive optics system for compensation of primary aberrations by software analysis of a point source image, Part II: tests, Applied Optics Vol. 46, No. 25, pp. 6427-643, 0003-6935, (2007).

[30] Judson R.S., Rabitz H., Teching lasers to control molecules, Phys. Rev. Lett., Vol. 68, No. 10, pp. 1079-7114 (1992).

[31] Minozzi M., Bonora S., Vallone G., Segienko A., Villoresi P., Bi-photon generation with optimized wavefront by means of Adaptive Optics, 11th International Conference on quantum communication, Vienna, Austria, 30 July-3 August, 2012.

[32] Bonora S., Distributed actuators deformable mirror for adaptive optics, Optics Communications, Vol. 284, No. 13, pp. 0030-4018 , (2011).

[33] Bonora S., Capraro I., Poletto L., Romanin M., Trestino C., Villoresi P., Fast wavefront active control by a simple DSP-Driven deformable mirror, Review of Scientific Instruments, Vol. 77, No. 9, pp. 0034-6748 ,(2006).

[34] Bortolozzo U., Bonora S., Huignard J.P., Residori S., Continuous photocontrolled deformable membrane mirror, Applied Physics Letters, Vol. 96, No.25, pp. 0003-6951, (2010).

[35] Bonora S., Coburn D., Bortolozzo U., Dainty C., Residori S., High resolution wavefront correction with photocontrolled deformable mirror, Optics Express Vol. 20, No. 5, pp. 5178-5188, (2012).

[36] Grisan E., Frassetto F., Da Deppo V., Naletto G., Ruggeri A., No wavefront sensor adaptive optics system for compensation of primary aberrations by software analysis of a point source image. Part I: methods, Applied Optics, Vol. 46, No. 25, pp. 6434-6441, 0003-6935, (2007).

[37] Fercher AF, Hitzenberger CK. Optical coherence tomography. In: Progress in Optics, Vol. 44, Chapter 4, pp. 215-302, Wolf E. Editor, (Elsevier Science & Technology, 2002).

[38] Zhang Y, Roorda A. Evaluating the lateral resolution of the adaptive optics scanning laser ophthalmoscope, J. Biomed. Opt., Vol. 11, No. 1, pp. 014002, (2006).

[39] Pircher M., Zawadzki R.J. , Combining adaptive optics with optical coherence tomography: Unveiling the cellular structure of the human retina in vivo, Expert Review of Ophthalmology, Vol. 2, No. 6, pp. 1019-1035, (2007).

[40] Zawadzki R.J., Jones S.M., Pilli S., Balderas-Mata S., Kim D., Olivier S.S., Werner J.S., Integrated adaptive optics optical coherence tomography and adaptive optics scanning laser ophthalmoscope system for simultaneous cellular resolution in vivo retinal imaging, Biomed. Opt. Express Vol. 2, No. 6, pp. 1674-1686, (2011).

[41] Muller R. A., Buffington A., Real-time correction of atmospherically degraded tele-scope images through image sharpening, J. Opt. Soc. Am., Vol. 64, No. 9, pp. 1200–1210, (1974).

[42] Hofer H., Sredar N., Queener H., Li C., Porter J., Wavefront sensorless adaptive optics ophthalmoscopy in the human eye, Opt. Express Vol. 19, No. 14160-14171, pp. 14160-14171, (2011)

[43] Bonora S., Frassetto F., Coraggia S., Spezzani C., Coreno M., Negro M., Devetta M, Vozzi, C., Stagira, S. Poletto L., Optimization of low-order harmonic generation by exploitation of a deformable mirror, Applied Physics B, Vol. 106, No. 4, pp.905-909, (2011).

[44] Vozzi C., Calegari F., Benedetti E., Gasilov S., Sansone G., Cerullo G., Nisoli M., De Silvestri S., Stagira S., Millijoule-level phase-stabilized few-optical-cycle infrared parametric source, Opt. Lett., Vol. 32, No. 20, pp. 2957-2959 (2007).

Chapter 4

Modeling and Control of Deformable Membrane Mirrors

Thomas Ruppel

Additional information is available at the end of the chapter

1. Introduction

The use of deformable mirrors (DMs) in adaptive optics (AO) systems allows for compensation of various external and internal optical disturbances during image aquisition. For example, an astronomical telescope equipped with a fast deformable secondary mirror can compensate for atmospheric disturbances and wind shake of the telescope structure resulting in higher image resolution [1–4]. In microscopy, deformable mirrors allow to correct for aberrations caused by local variations of the refractive index of observed specimen. Especially confocal and multi-photon microscopes particularly benefit from the improved resolution for visualization of cellular structures and subcellular processes [5, 6]. In addition, results of applied adaptive optics for detection of eye diseases and in vitro retinal imaging on the cellular level show promising examination and treatment opportunities [7–9].

In many AO systems, the deformable mirror is assumed to have negligible dynamical characteristics in comparison to the dynamic disturbances compensated by the deformable mirror. Unfortunately, this assumption is not always valid and active shape control of deformable mirrors must be employed to enhance the dynamic properties of the deformable mirror. For example, adaptive secondary mirrors for the Multi Mirror Telescope (MMT), the Large Binocular Telescope (LBT), and the Very Large Telescope (VLT) with diameters around 1 m have their first natural resonant frequencies below 10 Hz. In order to be able to use these systems for compensation of atmospherical disturbances with typical frequencies up to 100 Hz, active shape control is employed pushing the bandwith of these DMs to 1 kHz [10–13].

With up to 1170 voice coil actuators and co-located capacitive position sensors, the new generation of continuous face-sheet deformable mirrors requires fast and precise shape control. Thereby, the main idea for robust control of the mirror surface is the use of distributed voice coil actuators in combination with local position sensing by capacitive sensors. The use of voice coil actuators allows for large stroke and exact positioning of the mirror while it is floating in a magnetic field. By changing the spatial properties of the magnetic field, the mirror shell can be deformed into a desired shape to compensate for optical aberrations measured by the AO system. Thereby, a low-contact bearing combined with little intrinsic damping of the mirror shell draws the need for adequate damping in closed loop operation. For the MMT deformable mirror, this problem is tackled by a 40 μm air-gap between the mirror shell and a reference plate behind the shell. The induced viscous damping is sufficient to operate the mirror within the designed specifications [11]. However, recent deformable mirrors for the LBT and VLT shall operate within a larger air-gap to provide more stroke. As a side effect, the requirement for a larger air-gap reduces the natural viscous damping and electronic damping is needed to achieve the same control bandwidth [14]. Therefore, the LBT and VLT type deformable secondary mirrors are controlled by local PD-control for each actuator/sensor pair in combination with feedforward force compensation [15]. Each actuator is driven by a dedicated local position controller running at 40-70 kHz. The shape command for the mirror unit is generated by a higher level wavefront control loop running at about 1 kHz. For small set-point changes of the mirror shell, this control concept is well-suited and has proven to be applicable in practice [16].

For high speed deformations over large amplitudes (e.g. chopping of deformable secondary mirrors), disadvantages of local PD-control must be considered. First of all, there is a shape-dependent stiffness and damping variation of the mirror shell. In particular, each deformation of the shell requires a specific amount of external force by the distributed actuators [11]. If local control instead of global control is used for position control of the shell, then robust and subsequently conservative controller design is necessary for all local control loops. Secondly, interaction of neighboring actuators has to be studied carefully. The control loop gains have to be chosen such that only little interaction with neighboring control loops is caused. Otherwise, the local control concept can lead to instability of the shell.

In order to further investigate practical concepts for DM shape control, there have been studies on MIMO optimal feedback control [17–19]. Certainly, the closed loop performance of a global optimal feedback control concept is superior to local PD-control. But still, the computational load of a global MIMO controller may not be suitable for large deformable mirrors with more than thousand actuators. Only in [17], the circular symmetry of the mirror shell is used to reduce the controllers complexity. Thereby, it is shown that symmetry can effectively reduce computational loads without loosing control performance. Although, even when symmetry is fully exploited, the computational effort for a global feedback control concept of future deformable mirrors is considerably high.

The most difficult task for control of deformable mirrors clearly is not the stabilization of the shell in a static shape, but changing the mirrors deflection in a predefined time and maintaining system stability. Instead of using a pure feedback controller for this task, model-based feedforward control concepts are proposed as in [20, 21]. In the past, the concept of model-based feedforward control has successfully been applied to various classes of dynamical systems. This technique is widely used in control practice as an extension of a feedback control loop to separately design tracking performance by the feedforward part and closed-loop stability and robustness by the feedback part. By using model-based feedforward control for shape control a deformable mirror, the feedback control loop is only needed to stabilize the shell in steady state and along precomputed trajectories. Thereby, the feedback loop can be designed to achieve high disturbance rejection. The computational load for feedforward control is comparably low because the feedforward signals may not be computed in the feedback loop frequency (40-70 kHz for the LBT and VLT type DMs), but only when a new set-point is commanded (about 1 kHz for the LBT and VLT type DMs).

Since typical deformable mirrors do not significantly change their dynamical behavior over time, model-based time-invariant feedforward control can efficiently reduce load on implemented feedback controllers. Studies on model-based feedforward control of large deformable mirrors show that either with poorly tuned feedback control or even without feedback control, high speed and high precision deformations of deformable mirrors can be achieved [20, 21].[1] The required dynamical model of the DM can be identified based on internal position measurements of excited DM actuators. In [22], a practical identification procedure for small scale membrane DMs with static interferometric measurements and dynamic measurements from a laser vibrometer are presented, additionally. It is shown that together with an identified dynamical model, feedforward control can be employed for small scale membrane DMs also, significantly improving the settling time of the membrane mirror.

In the following, shape control of small and large deformable membrane mirrors by model-based feedforward and feedback control in a two-degree-of-freedom structure are described in a generalized framework. For this purpose, a scalable physical model of deformable membrane mirrors is derived based on force and momentum equilibriums of a differential plate element in Section 2. A series solution of the resulting homogeneous partial differential equation (PDE) is used to derive the modal coordinates of the inhomogeneous PDE including external actuator forces. This series solution can be employed to analyze and simulate the spatio-temporal behaviour of deformable mirrors and allows the design of a model-based shape controller in Section 4. The controller design section is devided into three parts. In Section 4.1, a trajectory generator for computation of differentiable reference trajectories describing the transient set-point change of the deformable mirror surface is described. In Section 4.2, a static and a dynamic feedforward controller are deduced from the inverse system dynamics of the mirror model, aftwards. Finally, a feedback controller is designed in Section 4.3 as a linear quadratic regulator based on a reduced dynamical mirror model in modal coordinates. After transforming the modal feedback controller into physical coordinates, its decentralized structure is implemented and its usability for future deformable mirrors is discussed.

2. Mirror modeling

Deformable membrane mirrors with non-contacting actuators are often represented by static models approximating the nonreactive deformation of the mirror surface. Thereby, both Kirchoff and van Kármán theory is used to describe plate deformations smaller than the plate thickness [23–26] and deformations close to the thickness of the mirror plate [27–29], respectively. Additionally, finite element methods are used to model deformable mirrors, in particular for the development large deformable secondary mirrors in astronomy [30–36]. In order to describe actuator influence functions, radially symmetric Gauss functions or splines are commonly used [2, 37], also. A more detailed static analysis is performed in [38] and [39] for a circularly clamped deformable mirror using a Kirchoff plate model. Additionally, in

---

[1] In this context, feedback control is mainly needed to account for model uncertainties and to reject external disturbances (e.g. mechanical vibrations, wind loads).

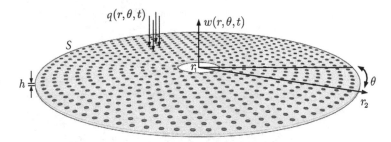

Figure 1. Schematics of a circular deformable mirror with deflection $w(r, \theta, t)$, inner radius r_1, outer radius r_2, plate thickness h, and external forces $q(r, \theta, t)$ acting on marked actuator positions over the mirror area S

[40] a deformable mirror with a free outer edge is modeled based on a Kirchoff plate model and resulting actuator influence functions for point forces are given.

Unfortunately, many of these modeling approaches only concentrate on the static characteristics of the deformable mirror and neglect dynamic properties like temporal eigenfrequencies and spatial characteristics of inherent eigenmodes. For model-based controller design addressed here, particularly these eigenfrequencies and eigenmodes are of substantial interest and can be found with the following modeling approach.

At first, circular deformable membrane mirrors with considerable out-of-plane stiffness and possible finite inner radius r_1 and outer radius r_2 centered at the origin of the $r - \theta$ plane are considered. Hereby, r, θ are polar coordinates. The plate is modelled as an isotropic Kirchoff plate with constant thickness h as shown in Figure 1.

For dynamic analysis, large deflections of the plate are not considered and nonlinear effects such as tensile stresses of the plate are neglected [28, 41, 42]. Secondly, it is assumed that a native curvature of the shell can be neglected due to only small deflections perpendicular to the surface.[2] The time varying deflection of the plate is measured by $w(r, \theta, t)$ relative to the undeflected reference and $t \in \mathbb{R}^+$ is the time. External forces caused by non-contacting voice coil actuators are represented by $q(r, \theta, t)$ and are defined by

$$q(r, \theta, t) = \sum_{m=1}^{M} \frac{1}{r_m} u_m(t) \delta(r - r_m) \delta(\theta - \theta_m), \tag{1}$$

where $u_m(t)$ corresponds to the actuator force at position (r_m, θ_m) and $\delta(r - r_m)\delta(\theta - \theta_m)$ describe the point-shaped force transmission.

The governing forces for a differential element of the plate are derived in Carthesian coordinates for simplicity. The differential element $hdxdy$ is effected by various shear forces, bending and twisting moments, and external loads as illustrated in Figure 2. The bending moments per unit length M_x, M_y arise from distributions of normal stresses σ_x, σ_y, while the twisting moments per unit length M_{xy}, M_{yx} (shown as double-arrow vectors) arise from

---

[2] As discussed in [43], a possibly large radius of curvature and a large diameter of the deformable mirror spherical face-sheet (e.g. $2m$) in comparison to the considered deflections (usually around $100\mu m$) allows to approximate the spherical shell by a Kirchhoff plate. Similar assumptions are drawn in [44] for a comparable system.

shearing stresses τ_{xy}, τ_{yx}. The shear forces per unit length Q_x, Q_y arise from shearing stresses τ_{xz}, τ_{yz} [41].

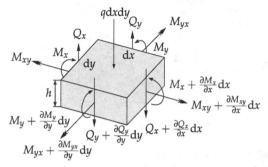

Figure 2. Illustration of shear forces, bending and twisting moments, and external loads affecting a differential plate element $hdxdy$ of the deformable mirror.

Of particular interest are the resulting three equations of motion

$$-Q_x dy + \left(Q_x + \frac{\partial Q_x}{\partial x}dx\right)dy - Q_y dx + \left(Q_y + \frac{\partial Q_y}{\partial y}dy\right)dx + qdxdy = \rho h dxdy \frac{\partial^2 w}{\partial t^2},$$
(2a)

$$\left(M_y + \frac{\partial M_y}{\partial y}dx\right)dx - M_y dx + M_{xy}dy - \left(M_{xy} + \frac{\partial M_{xy}}{\partial x}dx\right)dy - Q_y dxdy = 0,$$
(2b)

$$\left(M_x + \frac{\partial M_x}{\partial x}dx\right)dy - M_x dy - M_{yx}dx + \left(M_{yx} + \frac{\partial M_{yx}}{\partial y}dy\right)dx - Q_x dxdy = 0,$$
(2c)

with $x, y, t \in \mathbb{R}$. Rotary-inertia effects of plate elements as well as higher-order contributions to the moments from loading q have been neglected in the moment equations (2b) and (2c). By canceling terms, the equations of motion reduce to

$$\frac{\partial Q_x}{\partial x} + \frac{\partial Q_y}{\partial y} + q = \rho h \frac{\partial^2 w}{\partial t^2},$$
(3a)

$$\frac{\partial M_y}{\partial y} - \frac{\partial M_{xy}}{\partial x} - Q_y = 0,$$
(3b)

$$\frac{\partial M_x}{\partial y} + \frac{\partial M_{yx}}{\partial x} - Q_x = 0.$$
(3c)

By solving the last two equations for Q_x and Q_y and substituting in the first equation, a single equation in terms of various moments can be achieved reading

$$\frac{\partial^2 M_x}{\partial x^2} + \frac{\partial^2 M_{xy}}{\partial x \partial y} - \frac{\partial^2 M_{xy}}{\partial y \partial x} + \frac{\partial^2 M_y}{\partial y^2} + q = \rho h \frac{\partial^2 w}{\partial t^2}.$$
(4)

In [41], the relationship between the moments and the deflection is described and it is shown that the bending moments per unit length read

$$M_x = -D\left(\frac{\partial^2 w}{\partial x^2} + v\frac{\partial^2 w}{\partial y^2}\right), \tag{5a}$$

$$M_y = -D\left(\frac{\partial^2 w}{\partial y^2} + v\frac{\partial^2 w}{\partial x^2}\right), \tag{5b}$$

$$M_{xy} = -M_{yx} = D(1-v)\frac{\partial^2 w}{\partial x \partial y}, \tag{5c}$$

$$D = \frac{Eh^3}{12(1-v^2)}, \tag{5d}$$

with Young's modulus E and Poisson's ratio v. By using the relationship $M_{xy} = -M_{yx}$ the biharmonic partial differentail equation (PDE)

$$D\left(\frac{\partial^4 w}{\partial x^4} + 2\frac{\partial^4 w}{\partial x^2 \partial y^2} + \frac{\partial^4 w}{\partial y^4}\right) - q = -\rho h\frac{\partial^2 w}{\partial t^2} \tag{6}$$

is derived.

Using (1) for a point actuated circular plate, Equation (6) can be written in polar coordinates as

$$D\nabla^4 w(r,\theta,t) + (\lambda_d + \kappa_d\nabla^4)\frac{\partial w(r,\theta,t)}{\partial t} + \rho h\frac{\partial^2 w(r,\theta,t)}{\partial t^2} = \sum_{m=1}^{M}\frac{u_m(t)}{r_m}\delta(r - r_m)\delta(\theta - \theta_m), \tag{7}$$

with the biharmonic operator ∇^4 given by

$$\nabla^4 = \nabla^2\nabla^2 = \left(\frac{\partial^2}{\partial r^2} + \frac{1}{r}\frac{\partial}{\partial r} + \frac{1}{r^2}\frac{\partial^2}{\partial\theta^2}\right)^2. \tag{8}$$

The parameters λ_d and κ_d are used to characterize additionally included viscous and Rayleigh damping and need to be identified in practice.

In order to fully describe the spatio-temporal behavior of deformable mirrors, the biharmonic equation (7) must be completed by physically motivated boundary conditions of the mirror plate. Typical boundary conditions for deformable mirrors are illustrated in Figure 3 (a)-(c) and can either be a clamped edge at $r = r_1$ reading

$$w(r_1,\theta,t) = 0 \tag{9a}$$

$$\left.\frac{\partial w(r,\theta,t)}{\partial r}\right|_{r=r_1} = 0, \tag{9b}$$

or a simply supported edge at $r = r_1$ reading

$$w(r_1, \theta, t) = 0 \tag{10a}$$

$$\frac{\nu}{r^2} \frac{\partial^2 w(r,\theta,t)}{\partial \theta^2} + \frac{\partial^2 w(r,\theta,t)}{\partial r^2} + \frac{\nu}{r} \frac{\partial w(r,\theta,t)}{\partial r}\bigg|_{r=r_1} = 0, \tag{10b}$$

or a free edge at $r = r_1$ reading

$$\frac{\nu}{r^2} \frac{\partial^2 w(r,\theta,t)}{\partial \theta^2} + \frac{\partial^2 w(r,\theta,t)}{\partial r^2} + \frac{\nu}{r} \frac{\partial w(r,\theta,t)}{\partial r}\bigg|_{r=r_1} = 0 \tag{11a}$$

$$\frac{(\nu-2)}{r^2} \frac{\partial^3 w(r,\theta,t)}{\partial r \partial \theta^2} - \frac{\partial^3 w(r,\theta,t)}{\partial r^3} + \frac{(3-\nu)}{r^3} \frac{\partial^2 w(r,\theta,t)}{\partial \theta^2}$$
$$- \frac{1}{r} \frac{\partial^2 w(r,\theta,t)}{\partial r^2} + \frac{1}{r^2} \frac{\partial w(r,\theta,t)}{\partial r}\bigg|_{r=r_1} = 0. \tag{11b}$$

In order to include a flexible support at $r = r_1$ (see Figure 3 (d)), boundary conditions (11) can be extended by a righting moment c reading

$$-c \frac{\partial w(r,\theta,t)}{\partial r} + \frac{\nu}{r^2} \frac{\partial^2 w(r,\theta,t)}{\partial \theta^2} + \frac{\partial^2 w(r,\theta,t)}{\partial r^2} + \frac{\nu}{r} \frac{\partial w(r,\theta,t)}{\partial r}\bigg|_{r=r_1} = 0, \tag{12a}$$

$$\frac{(\nu-2)}{r^2} \frac{\partial^3 w(r,\theta,t)}{\partial r \partial \theta^2} - \frac{\partial^3 w(r,\theta,t)}{\partial r^3} + \frac{(3-\nu)}{r^3} \frac{\partial^2 w(r,\theta,t)}{\partial \theta^2}$$
$$- \frac{1}{r} \frac{\partial^2 w(r,\theta,t)}{\partial r^2} + \frac{1}{r^2} \frac{\partial w(r,\theta,t)}{\partial r}\bigg|_{r=r_1} = 0. \tag{12b}$$

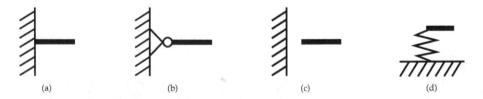

Figure 3. Schematics of four typical boundary conditions for deformable membrane mirrors showing (a) a clamped edge, (b) a simply supported edge, (c) a free edge, and (c) a spring supported edge type.

The PDE (7) and the boundary conditions (9)-(11) fully describe both the temporal and the spatial behavior of the modelled deformable mirror. In order to quantify the resulting eigenfrequencies and eigenfunctions for a given mirror geometry and material properties, the PDE will be analyzed further in the following section.

3. Fundamental solution and modal analysis

One way to analyze PDE (7) is to discretize the PDE in its spatial coordinates (r, θ). Thereby, the spatial differential operator ∇^4 and the boundary conditions (9)-(11) can be approximated by finite difference or finite element methods and the PDE is reduced to a finite set of coupled ordinary differential equations (ODE). The Eigenfrequencies and the eigenvectors of the resulting generalized eigenvalue problem correspond to the spatially discretized eigenfunctions of the PDE. Thereby, the spatial discretization has strong influence on the remaining temporal system dynamics and can even mislead to wrong results if performed insufficiently. The typically large set of ordinary differential equations requires high computing power for dynamic simulation and is ill-suited for model-based controller design, in general.

Another way to analyze PDE (7) is a modal transformation based on the eigenfunctions of PDE (7) fulfilling the homogeneous boundary conditions (9)-(11). The modal transformation leads to an infinite set of ordinary differential equations in modal coordinates describing only the temporal evolution of the eigenfunctions of PDE (7). Thereby, all spatial and temporal properties are preserved and a reduced set of the ODEs can be used for dynamic simulation and controller design.

The eigenfuntions $W_k(r, \theta)$ of PDE (7) can be derived by separation of variables with

$$\Gamma: \quad w(r, \theta, t) = \sum_{k=1}^{\infty} W_k(r, \theta) f_k(t), \tag{13}$$

where the eigenfunctions $W_k(r, \theta)$ describe the spatial characteristics and the modal coefficients $f_k(t)$ describe the time-varying amplitude. At the same time, equation (13) can be used to perform a transformation from modal coordiantes $f_k(t)$ to physcial coordinates $w(r, \theta, t)$ and back, which will be needed for feedback controller design in Section 4.3.

After inserting (13) in the homogeneous PDE (7) with $q(r, \theta) = 0$, an analytical description of the eigefunctions $W_k(r, \theta, t)$ can be found as described in [23] reading

$$W_k(r, \theta) = (A_{1k} J_k(\beta r) + A_{2k} Y_k(\beta r) + A_{3k} I_k(\beta r) + A_{4k} K_k(\beta r)) \cos(k\theta)$$
$$+ (B_{1k} J_k(\beta r) + B_{2k} Y_k(\beta r) + B_{3k} I_k(\beta r) + B_{4k} K_k(\beta r)) \sin(k\theta). \tag{14}$$

Here, J_k, Y_k, I_k, K_k are Bessel functions and modified Bessel functions of first and second kind. The eigenfunctions fulfill the relation

$$\nabla^4 W_k(r, \theta) = \beta^4 W_k(r, \theta) \tag{15}$$

and are self-adjoint due to the biharmonic operator ∇^4.

After inserting the fundamental solution (14) in the boundary conditions (9)-(11), the free parameters $A_{1k}, \ldots, A_{4k}$ and $B_{1k}, \ldots, B_{4k}$ can be found by computing a non-trivial solution of the resulting homogeneous equations

$$\Lambda_c(\beta_k) \begin{bmatrix} A_{1k} \\ A_{2k} \\ A_{3k} \\ A_{4k} \end{bmatrix} = 0, \quad \Lambda_s(\beta_k) \begin{bmatrix} B_{1k} \\ B_{2k} \\ B_{3k} \\ B_{4k} \end{bmatrix} = 0, \tag{16}$$

where $\Lambda_c(\beta)$ and $\Lambda_s(\beta)$ are $[4 \times 4]$ matrices containing linear combinations of Bessel functions depending on the eigenvalue β_k for the cosine and sine parts of $W_k(r,\theta)$. A computation of matrices Λ_c and Λ_s should be performed using computer algebra software since the matrix elements are rather extensive.

where $\Lambda_c(\beta)$ and $\Lambda_s(\beta)$ are $[4 \times 4]$ matrices containing linear combinations of Bessel functions depending on the eigenvalue β_k for the cosine and sine parts of $W_k(r,\theta)$. A computation of matrices Λ_c and Λ_s should be performed using computer algebra software since the matrix elements are rather extensive.

In order to compute a non-trivial solution of the homogeneous equations (16), the parameter β_k needs to determined such that Λ_c and Λ_s contain a non-trivial kernel. This can be achieved by numerically searching for zeros in the determinant of the matrices Λ_c and Λ_s and leads to infinitely many eigenvalues β_k related to the eigenfunction $W_k(r,\theta)$.

In Figure 4, the consine parts of the first 21 analytical eigenfunctions $W_k(r,\theta)$, $k = 1,\ldots,21$ of a deformable mirror with clamped inner radius and free outer radius are shown based on the calculations given before. The eigenfunctions $W_k(r,\theta)$ and eigenvalues β_k were compared with a detailed finite element model of the same mirror and show excellent compliance.

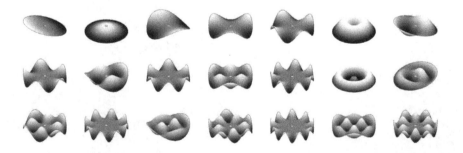

Figure 4. First 21 eigenfunctions of a deformable mirror model (7) with clamped edge boundary conditions (9) at the inner radius r_1 and free edge boundary conditions (11) at the outer radius r_2 ordered by increasing eigenfrequency.

In the following, a modal decomposition of PDE (7) is used to analyze the mirror dynamics and to derive model-based feedforward control commands in modal coordinates. The modal decomposition of (7) is performed by inserting relation (13) in (7) and using property (15) yielding

$$\sum_{k=1}^{\infty} \left(D\beta_k^4 f_k(t) + (\lambda_d + \kappa_d \beta^4)\dot{f}_k(t) + \rho h \ddot{f}_k(t) \right) W_k(r,\theta) = \sum_{m=1}^{M} \frac{u_m(t)}{r_m} \delta(r - r_m)\delta(\theta - \theta_m). \quad (17)$$

A multiplication with any eigenfunction W_j and integration over domain S on both sides gives

$$\iint_S W_j(r,\theta) \sum_{k=1}^{\infty} \left(D\beta_k^4 f_k(t) + (\lambda_d + \kappa_d \beta_k^4)\dot{f}_k(t) + \rho h \ddot{f}_k(t) \right) W_k(r,\theta)\,dS =$$

$$\iint_S W_j(r,\theta) \sum_{m=1}^{M} \frac{u_m(t)}{r_m} \delta(r - r_m)\delta(\theta - \theta_m)\,dS.$$

(18)

Changing order of summation and integration, using the orthonormality property of the eigenfunctions $W_k(r,\theta)$, and considering the sifting property of the Dirac delta function δ on the right hand side of Equation (18) leads to an infinite set of second order ordinary differential equations

$$\ddot{f}_j(t) + \frac{\lambda_d + \kappa_d \beta_j^4}{\rho h}\dot{f}_j(t) + \frac{D\beta_j^4}{\rho h}f_j(t) = \frac{1}{\rho h}\sum_{m=1}^{M} W_j(r_m,\theta_m)u_m(t), \quad (19)$$

$$f_j(0) = 0, \quad \dot{f}_j(0) = 0, \qquad j \in \mathbb{N}.$$

For each j, the resulting ODE (19) describes the temporal evolution of the corresponding eigenfunction $W_j(r,\theta)$ by the modal coefficient $f_j(t)$. For simplicity, homogeneous initial conditions are assumed describing a flat and steady mirror surface.

Since there is no coupling between different modal coefficients $f_k(t)$ and $f_j(t)$, $j \neq k$, the ODEs (19) can be used for dynamic simulation of a mirror surface described by a specific eigenfunction $W_j(r,\theta)$ or a linear combination of eigenfunctions. Any physically relevant mirror deformation can be decomposed into a linear combination of the orthonormal eigenfunctions $W_k(r,\theta)$ and all relevant dynamic properties of the deformable mirror can be covered with this approach. The decoupled description of the mirror dynamics in modal coordinates (19) allows for simplified system analysis and is the foundation for the following controller design.

Based on the derived normalized eigenfunctions $W_k(r,\theta)$, an analytical solution of the biharmonic equation (7) with initial conditions $w(r,\theta,0) = 0$ and $\dot{w}(r,\theta,0) = 0$ can be derived in spectral form reading

$$w(r,\theta,t) = \int\limits_{\tau=0}^{t} \underbrace{\sum_{k=1}^{\infty} \sum_{m=1}^{M} \frac{1}{\zeta_k} W_k(r,\theta) \, W_k(r_m,\theta_m) g_k(t-\tau)}_{G(r,\theta,r_m,\theta_m,t-\tau)} u_m(\tau)\mathrm{d}\tau, \tag{20}$$

with

$$g_k(t-\tau) = e^{-\frac{t}{2\rho h}\left(\lambda_d + \kappa_d \beta_k^4 + \zeta_k\right)} - e^{-\frac{t}{2\rho h}\left(\lambda_d + \kappa_d \beta_k^4 - \zeta_k\right)}, \quad \zeta_k = \sqrt{(\lambda_d + \kappa_d \beta_k^4)^2 - 4D\rho h \beta_k^4}.$$

Equation (20) can be used to compute the actuator influence functions for any input signal $u_m(\tau)$, $m = 1 \ldots M$. Thereby, not only the resulting static deformation of the mirror can be computed, but also the transient motion of the plate for time-varying input forces.[3] Additionally, relation (20) can be used to compute the frequency responses of a deformable mirror at different actuator locations as shown in Figure 5 for a deformable mirror with clamped inner edge, free outer edge, and co-located force actuators and position sensors.

The infinite sum of eigenfunctions $W_k(r,\theta)$ in Equation (20) can be approximated with a finite k using only a limited number of eigenfunctions. The right number of eigenfunctions can either be driven by a sufficient static coverage of mirror deformations with linear combinations of k eigenfunctions $W_k(r,\theta)$, or by considering a certain number of eigenvalues β_k in order to sufficiently describe the mirror dynamics up to a certain eigenfrequency.

Obviously, the frequency responses shown in Figure 5 reveal essential variations of local mirror dynamics depending on the co-located actuator/sensor position. From a control point of view, this behavior illustrates the difficulty of designing a decentralized controller that can be used for every actuator/sensor pair of the DM.

4. Model-based controller design

In a typical AO system, the requirement of changing the mirror shape within a predefined time T from an initial deformation $\triangle w_1^*$ to a final deformation $\triangle w_2^*$ is derived from the higher-ranking AO control loop (see Figure 6). This control loop runs at a fixed cycle time and sends mirror deformations $\triangle w^*(t)$ to the shape controller in order to correct for measured optical disturbances in the AO system (not shown in Figure 6). The step inputs $\triangle w^*(t)$ for the model-based shape controller in Figure 6 are in general ill-suited for model-based feedforward control and need to be filtered beforehand. The transition time T for changing the mirror shape from its initial shape $\triangle w_1$ to its new desired shape $\triangle w_2$ is assumed to be smaller than the cycle time of the higher-ranking AO control loop. This requirement assures that the closed loop mirror dynamics can be neglected with respect to the higher-ranking AO control loop when designing the AO loop controller.

For model-based shape control of deformable mirrors, the control scheme shown in Figure 6 can be used. It consists of a deformable mirror with multiple inputs $u = \begin{bmatrix} u_1 \, u_2 \, \ldots \, u_M \end{bmatrix}$

---

[3] In Equation (20) and for the following analysis, it is assumed that the eigenfunctions $W_k(r,\theta)$ are normalized with respect to the L_2 scalar product $\langle W_k, W_k \rangle = 1$.

(force actuators), multiple outputs $y = [y_1 \; y_2 \; \ldots \; y_M]$ (position sensors), and an underlying shape controller in a two-degree-of-freedom control structure. The control structure contains an online trajectory generator Σ^*_{traj}, a static and a dynamic model-based feedforward controller Σ^{-1}_{stat} and Σ^{-1}_{dyn}, and a model-based feedback controller Σ_{crtl}.

There are two control objectives adressed by the control structure shown in Figure 6. One is achieving good tracking performance along a spatio-temporal trajectory y_d describing the evolution of the mirror shape from an initial shape $\triangle w^*_1$ to a new commanded mirror shape $\triangle w^*_2$ in a predefined time T. The second control objective is the stabilization of the deformable mirror shell by feedback control Σ_{ctrl} along the generated trajectory y_d and in its final position $\triangle w^*_2$ with respect to external disturbances and model uncertainties.

4.1. Design of the trajectory generator Σ^*_{traj}

The trajectory generator Σ^*_{traj} generates a continuously differentiable output signal y_d that reaches its final value $\triangle w^*_2$ within a fixed transition time T. In Figure 7, the piecewise constant input $\triangle w^*$ and filtered output signal y_d are illustrated for a single set-point change. Since the deformable mirror consists of many inputs and outputs, the signals $\triangle w^*$ and y_d are vectorial variables containing deformation values either in modal or physical coordinates. The choice of units can be driven by the higher-ranking AO control loop and is only of minor importance for the following section.[4]

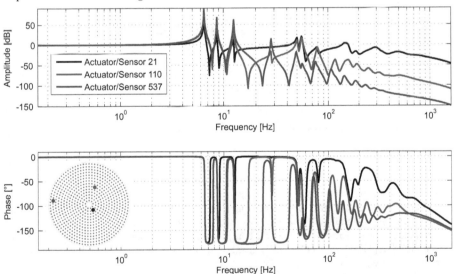

Figure 5. Bode diagram of the analytical local transfer functions of a deformable mirror at different co-located actuator/sensor positions (normalized).

---

[4] A linear forward or backward transformation can be used to transform physical coordinates $w(r, \theta, t)$ into modal coordinates $f_k(t)$ using the orthonormal eigenfunctions $W_k(r, \theta)$ in 13.

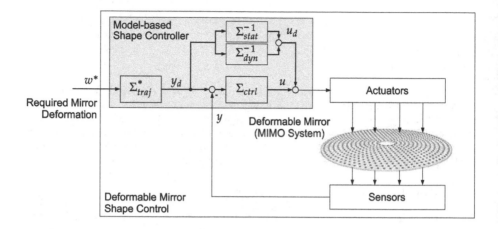

Figure 6. Two-degree-of-freedom structure for shape control of deformable mirrors modeled as a multi-input multi-ouput (MIMO) system with co-located non-contacting force actuators and position sensors consisting of a trajectory generator Σ^*_{traj}, a static and a dynamic model-based feedforward controller Σ^{-1}_{stat} and Σ^{-1}_{dyn}, and a feedback controller Σ_{ctrl}.

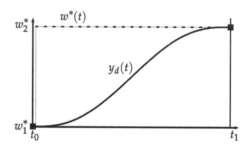

Figure 7. Illustration of a general input signal $\triangle w^*(t)$ and the resulting output signal $y_d(t)$ of the trajectory generator Σ^*_{traj} within a fixed transition time $T = t_2 - t_1$.

There are several ways to generate the output signal y_d based on the step input $\triangle w^*$, e.g. finite impuls response (FIR) filters, sigmoid basis functions, or segmented polynomials. In [20] and [21] a detailed introduction to this topic is given with respect to control of deformable mirrors. A more general and comprehensive discussion on trajectory generation can be found in [45]. Here, a polynomial ansatz function is used to generate the output signal y_d for simplicity.

Depending on the required smoothness of y_d, the reference trajectory y_d is described by

$$\Sigma^*_{traj}: \quad y_d(t) = \triangle w^*_1 + (\triangle w^*_2 - \triangle w^*_1)\kappa(t - t_0), \quad t \in [t_0, t_1], \; t_1 = t_0 + T, \quad (21)$$

with

$$
\kappa(t) = \begin{cases} 0 & \text{if } t \leq (t_1 - t_0) \\ \frac{(2n+1)!}{n!(t_1-t_0)^{2n+1}} \sum_{i=0}^{n} \frac{(-1)^{n-i}}{i!(n-i)!(2n-i+1)} (t_1 - t_0)^i t^{2n-i+1} \\ & \text{if } 0 \leq t \leq (t_1 - t_0) \\ 1 & \text{if } t \geq (t_1 - t_0). \end{cases} \tag{22}
$$

The transition time T must be chosen shorter than the higher-ranking AO loop cycle time but long enough to comply with input constraints of the force actuators. The shorter the transition time T is chosen, the higher the required input forces u will be in order to drive the mirror from one deformation to another one. The required smoothness of the trajectory y_d depends on the parameter n and should be chosen $n \geq 3$ in order to be able to apply the model-based feedforward control scheme presented in the next section.

4.2. Design of the model-based feedforward controller Σ_{stat}^{-1} and Σ_{dyn}^{-1}

The basic idea of model-based feedforward control is the application of a desired input signal $u_d(t)$ such that the system response $y(t)$ follows a desired output behavior $y_d(t)$ (see Figure 6). The transient deformation of the deformable mirror is described by the planned output $y_d(t)$ and leads to a new steady shape $\triangle w_2^*$ using an inverse model of the system dynamics of the DM. The design of the desired response $y_d(t)$ is performed in the on-line trajectory generator Σ_{traj}^* described in Section 4.1.

As shown in Figure 6, the feedforward control concept is seperated into a static and a dynamic component Σ_{stat}^{-1} and Σ_{dyn}^{-1}. The feedforward component Σ_{dyn}^{-1} contains the dynamic inverse of the relevant mirror dynamics and is based on the modal system dynamics (19) reading

$$
\ddot{f}_j(t) + \underbrace{\frac{\lambda_d + \kappa_d \beta_j^4}{\rho h}}_{2d_j\omega_j} \dot{f}_j(t) + \underbrace{\frac{D\beta_j^4}{\rho h}}_{\omega_j^2} f_j(t) = \underbrace{\frac{1}{\rho h} \sum_{m=1}^{M} W_j(r_m, \theta_m) u_m(t)}_{b_j}, \tag{23}
$$

By transforming the desired output behavior y_d into modal coordinates $f_j(t)$ with relation (13), the left hand side of Equation (23) is fully defined by the desired modal output behavior $f_k(t)$ and the first two time derivatives of $f_k(t)$. Thus, the dynamic part of the required input command u_d achieving the planned output behavior $f_k(t)$ can be computed as

$$
\Sigma_{dyn}^{-1}: \quad u_m^{dyn}(t) = \frac{1}{b_j} \left(\ddot{f}_j(t) + 2d_j\omega_j\dot{f}_j(t) + \omega_j^2 f_j(t) \right) \tag{24}
$$

for a limited number of $j = 1, \ldots, N$ dynamic eigenmodes. Since viscous and Rayleigh damping typically increase for higher order eigenmodes, it is not necessary to consider all eigenmodes in the dynamic feedforward component Σ_{dyn}^{-1}. Instead, it is sufficient to

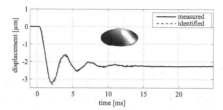 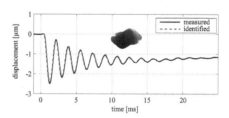

Figure 8. Measured step responses of the first and second eigenfunctions of an Alpao DM88 deformable membrane mirror without feedforward control showing a settling time above 5-10 ms

consider only the static components of the inverse system dynamics (23) and compute a static command $u_m^{stat}(t)$ as

$$\Sigma_{stat}^{-1} : \quad u_m^{stat}(t) = \frac{1}{b_j}\left(\omega_j^2 f_j(t)\right). \tag{25}$$

Afterwards, the final feedforward command u_d can be computed as

$$u_d(t) = u^{dyn}(t) + u^{stat}(t). \tag{26}$$

Regarding the computational complexity, in [20] is shown that the computational demands of feedforward control scale quadratically with the number of system inputs and outputs. However, since the control signal only needs to be computed with the frequency of the outer AO-loop (e.g. 1 kHz in modern astronomical AO systems), there are currently no burdens for a practical implementation in existing systems.

In Figure 8 and Figure 9 measurement results of a setpoint change without and with feedforward control for an Alpao deformable mirror with 88 voice coil actuators is shown. In the experiment, the first two eigenfunctions of the deformable mirror are excited by a modal step input (spatially distributed step command for all 88 actuators at the same time) and a model-based feedforward control input u_d (spatially distributed time-varying signal for all 88 actuators) based on a polynomial of degree $n = 3$ and identified modal damping and eigenfrequencies in (24). Clearly, overshoot and settling time of the feedforward scheme show considerable improvements to a pure step command. Thereby, no feedback control is implemented at this stage and only the trajectory generator Σ_{traj}^* and the inverse modal system dynamics Σ_{dyn}^{-1} are used to generate the control command $u = u_d$ (see Figure 6).

Feedforward control of deformable membrane mirrors has also been demonstrated at the P45 adaptive secondary prototype of the Large Binocular Telescope adaptive secondary mirror in [21] with similar results. In addition, a flatness based feedforward control is proposed therein in case of zero dynamics in the differential equation (19).

In order to further improve the system response to feedforward control and in order to add additional disturbance rejection capabilities to the DM control scheme, a model-based

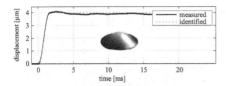 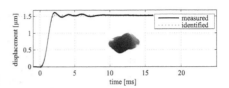

Figure 9. Measured responses of the first and second eigenfunctions of an Alpao DM88 deformable membrane mirror with feedforward control showing a settling time of 2 ms

feedback controller is designed in the following section and its sparse structure for local implementation is discussed.

4.3. Design of the model-based feedback controller Σ_{ctrl}

Feedforward control combined with suitable trajectory generation methods improve the input output behavior of deformable mirrors significantly. However, in case of model uncertainties or external disturbances, the tracking performance shown in Figure 9 can be affected considerably. For this reason, active position feedback is integrated in modern deformable mirrors measuring the local mirror position at each actuator generating an error compensation command.

Due to the spatially distributed mirror dynamics, global instead of local position control of deformable mirrors is the most promising solution for error compensation and disturbance rejection. However, the computational complexity for high order spatial control of the deformable element typically exceeds available computing power. For this reason, existing deformable membrane mirrors for large telescopes incorporate local feedback instead of global feedback control and neglect some of the global dynamics of the deformable mirror [10, 13, 14, 46–48]. As a side effect, dynamic coupling of separately controlled actuators through the deformable membrane can lead to instability of the individually stable loops and draws the need for carefully designing the control parameters of the local feedback loops.

In the following, an advanced control concept for position control of large deformable mirrors is derived based on the detailed dynamical model of the deformable mirror (7), suitable boundary conditions (9)-(12), and its represenation in modal coordinates (23). The presented controller design differs from existing ones since it incorporates a detailed mirror model and comprises a decentralized structure at the same time.

For feedback controller design, the first N relevant modal differential equations (23) are combined in state space form as

$$\dot{x}(t) = Ax(t) + Bu(t)$$
$$y(t) = Cx(t), \tag{27}$$

with $x(t) = \begin{bmatrix} f_1(t) & f_2(t) & \dots & f_N(t) & \dot{f}_1 & \dot{f}_2 & \dots & \dot{f}_N(t) \end{bmatrix}^T$ and system matrices

$$A = \begin{bmatrix}
0 & 0 & 0 & \dots & 0 & 1 & 0 & 0 & \dots & 0 \\
\frac{D\beta_1^4}{\rho h} & 0 & 0 & \dots & 0 & \frac{\lambda_d + \kappa_d \beta_1^4}{\rho h} & 0 & 0 & \dots & 0 \\
0 & 0 & 0 & \dots & 0 & 0 & 1 & 0 & \dots & 0 \\
0 & \frac{D\beta_2^4}{\rho h} & 0 & \dots & 0 & 0 & \frac{\lambda_d + \kappa_d \beta_2^4}{\rho h} & 0 & \dots & 0 \\
0 & 0 & 0 & \dots & 0 & 0 & 0 & 1 & \dots & 0 \\
0 & 0 & \frac{D\beta_3^4}{\rho h} & \dots & 0 & 0 & 0 & \frac{\lambda_d + \kappa_d \beta_3^4}{\rho h} & \dots & 0 \\
\vdots & \vdots & \vdots & \ddots & \vdots & \vdots & \vdots & \vdots & \ddots & \vdots \\
0 & 0 & 0 & \dots & 0 & 0 & 0 & 0 & \dots & 1 \\
0 & 0 & 0 & \dots & \frac{D\beta_N^4}{\rho h} & 0 & 0 & 0 & \dots & \frac{\lambda_d + \kappa_d \beta_N^4}{\rho h}
\end{bmatrix}, \qquad (28)$$

$$B = \frac{1}{\rho h} \begin{bmatrix}
W_1(r_1, \theta_1) & \dots & W_1(r_M, \theta_M) \\
W_2(r_1, \theta_1) & \dots & W_2(r_M, \theta_M) \\
\vdots & \dots & \vdots \\
W_N(r_1, \theta_1) & \dots & W_N(r_M, \theta_M) \\
0 & \dots & 0 \\
\vdots & \ddots & \vdots \\
0 & \dots & 0
\end{bmatrix}, \qquad (29)$$

$$C = \begin{bmatrix}
W_1(r_1, \theta_1) & \dots & W_N(r_1, \theta_1) & 0 \dots 0 \\
W_1(r_2, \theta_2) & \dots & W_N(r_2, \theta_2) & 0 \dots 0 \\
\vdots & \ddots & \vdots & \vdots \ddots \vdots \\
W_1(r_M, \theta_M) & \dots & W_N(r_M, \theta_M) & 0 \dots 0
\end{bmatrix}. \qquad (30)$$

A global multi-input multi-ouput linear quadratic regulator is designed for controlling the system along the planned trajectory $x_d(t)$ using the optimization criterion

$$J = \int_0^\infty (x(t) - x_d(t))^T Q (x(t) - x_d(t)) + u(t)^T R u(t) dt, \qquad (31)$$

with Q being a positive definite and R being a positive semi-definite matrix. Assuming the pair (A, Q) is observable, a minimization of (31) can be achieved by the state feedback

$$\Sigma_{ctrl}: \quad u(t) = -K (x(t) - x_d(t)) \qquad (32)$$

with

$$K = R^{-1}B^T P. \tag{33}$$

The matrix P is the symmetric, positive definite solution of the matrix Ricatti equation

$$A^T P + PA - PBR^{-1}B^T P + Q = 0 \tag{34}$$

and can be computed efficiently in modern mathematical computing languages (e.g. Matlab). By choosing $Q = C^T c_1 C$ and $R = Ic_2$, the optimization criterion for a deformable mirror with M inputs and M outputs simplifies to

$$J = \int_0^\infty \sum_{m=1}^M \left(c_1 y_m^2(t) + c_2 u_m^2(t) \right) dt \tag{35}$$

and a decentralized controller can be computed via (33) and (34). The particular choice of weight matrices Q and R is the essential step for a decentralized controller in the LQR framework. The coefficients c_1 and c_2 are used to tune the closed loop disturbance rejection and robustness until certain settling time and gain margins are achieved.

Since the controller K requires modal signals $x(t)$, a feedback controller K_p in physical coordinates can be computed performing an inverse modal transformation Γ^{-1} from (13) reading

$$K_p = \Gamma^{-1}K. \tag{36}$$

with the feedback law

$$u(t) = -K_p \left(\left[x(t) \ \dot{x}(t) \right]^T - \left[x_d(t) \ \dot{x}_d(t) \right]^T \right). \tag{37}$$

For comparison, the closed loop transfer functions at selected co-located actuator/sensor locations with active LQ feedback are shown in Figure 10. In comparison to the open loop transfer functions in Figure 5, all relevant low and high frequency resonances are fully damped in the closed loop case and the resulting bandwidth of the deformable mirror exceeds 1 kHz. In order to tune the resulting performance for existing deformable mirrors, a variation of parameters c_1 and c_2 can be performed easily in practice.

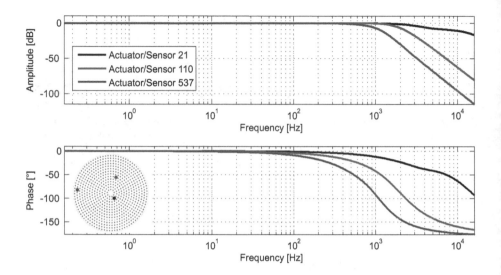

Figure 10. Bode diagram of the closed loop local transfer functions at different co-located actuator/sensor positions using decentralized LQR control (normalized).

In Equation (37), it can be seen that the columns of K_p contain information about how much displacement and velocity feedback is required for a certain actuator in vector $u(t)$. A visualization of entries in K_p shown in Figure 11 reveals that only a limited amount of displacement and velocity information around each actuator is needed to compute the feedback signal $u(t)$. This property can be used to truncate the spatial extension of the global LQ regulator and leads to a decentralized control scheme.

For a deformable mirror with 672 actuators and sensors, the corresponding normalized entries of the state feedback matrix K_p are visualized in Figure 11. Clearly, a choice of $c_1 = 1000$ and $c_2 = 0.1$ results in a fully decentralized structure of K_p. Comparing the position and velocity feedback entries of K_p in Figure 11, dynamic effects of boundary conditions and actuator position can be seen. Consequently, a model-free decentralized control law for all actuators seems unreasonable and the suggested linear quadratic regulator approach should be considered when designing deformable mirror controllers for astronomical telescopes or comparible application areas. Depending on the truncation area, the computational demands of this control concept scale linearly with the number of actuators and show the applicability of global LQ - control for shape control of large deformable mirrors in general.

Although controller K requires full state information, the state feedback controller can be transformed into an output feedback controller using loop transfer recovery (LTR) and results in an output controller Σ_{ctrl} as shown in Figure 6. The ouput controller can be implemented on existing hardware as a finite impulse response (FIR) filter for each actuator/sensor pair where the number of filter coefficients is mainly driven by the acceptable approximation error of the loop transfer recovery approach.

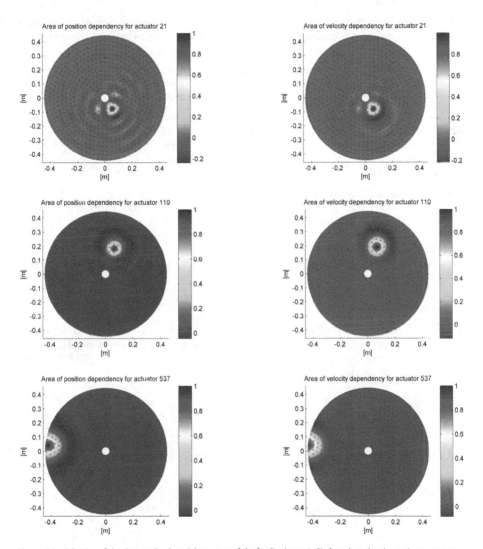

Figure 11. Vislization of the decentralized spatial structure of the feedback matrix K_p for selected co-located actuator/sensor pairs.

5. Conclusion

Control of deformable mirrors becomes relevant when the mirror dynamics are slower or equivalent with the dynamics of optical disturbances that shall be compensated by the deformable mirror. Active shape control of deformable mirrors can thereby increase the mirrors bandwidth and make it suitable for the AO task, again. When designing a model-based shape controller for a deformable mirror in a two-degree-of-freedom control structure, there are three components to be considered (see Figure 6): First, a trajectory generator Σ_{traj} providing continuous reference trajectories y_d for fast setpoint changes of the DM shape. Second, a static and dynamic feedforward controller Σ_{stat}^{-1} and Σ_{dyn}^{-1} generating control commands for driving the mirror along the precomputed trajectory based on inverse system dynamics. Finally, a feedback controller Σ_{ctrl} responsible for compensation of model errors in the feedforward part and rejection of external disturbances.

In this chapter, model-based design steps for all three components were shown supported by experimental and simulation results. The key for decentralized controller design is the right choice of weighting matrices Q and R in the LQR framework. Followed by a loop transfer recovery approach, the state feedback controller can be transformed into an ouput feedback controller that can be implemented as FIR filters on existing hardware.

Future astronomical optical telescopes, e.g. the European Extremely Large Telescope (E-ELT) or the Giant Magellan Telescope (GMT), will include deformable mirrors with many thousand actuators and position sensors. Due to the inherent slow dynamics of the large deformable mirror shells, active shape control of these elements is inevitable. Model-based shape control in a two-degree-of-freedom structure can greatly improve the performance of these elements and should be considered for comparible AO systems with high performance requirements, also.

Acknowledgements

The author would like to acknowledge the support by INAF and the Osservatorio Astrofisico di Arcetri in particular, providing finite element data of the LBT ASM and detailed insight into current deformable mirror control schemes. This work was supported by the DFG under grant SA-847/10-1 and OS-111/29-1 and by the Carl Zeiss AG.

Author details

Thomas Ruppel

Carl Zeiss AG, Corporate Research and Technology, Optronics Systems, Jena, Germany

References

[1] Edward P. Wallner. Optimal wave-front correction using slope measurements. *J. Opt. Soc. Am.*, 73(12):1771–1776, 1983.

[2] R.K. Tyson. *Adaptive optics engineering handbook.* CRC, 2000.

[3] B.W. Frazier and R.K. Tyson. Robust control of an adaptive optics system. In *System Theory, 2002. Proceedings of the Thirty-Fourth Southeastern Symposium on*, pages 293–296, 2002.

[4] H. M. Martin, G. Brusa Zappellini, B. Cuerden, S. M. Miller, A. Riccardi, and B. K. Smith. Deformable secondary mirrors for the LBT adaptive optics system. In Domenico Ellerbroek, Brent L.; Bonaccini Calia, editor, *Advances in Adaptive Optics II*, volume 6272 of *Proc. SPIE*, page 0U, July 2006.

[5] M.J. Booth, M.A.A. Neil, R. Juškaitis, and T. Wilson. Adaptive aberration correction in a confocal microscope. *Proceedings of the National Academy of Sciences of the United States of America*, 99(9):5788, 2002.

[6] M.J. Booth. Adaptive optics in microscopy. *Philosophical Transactions of the Royal Society A: Mathematical, Physical and Engineering Sciences*, 365(1861):2829, 2007.

[7] Junzhong Liang, David R. Williams, and Donald T. Miller. Supernormal vision and high-resolution retinal imaging through adaptive optics. *J. Opt. Soc. Am. A*, 14(11):2884–2892, 11 1997.

[8] Austin Roorda, Fernando Romero-Borja, III William Donnelly, Hope Queener, Thomas Hebert, and Melanie Campbell. Adaptive optics scanning laser ophthalmoscopy. *Opt. Express*, 10(9):405–412, 05 2002.

[9] Daniel C. Gray, William Merigan, Jessica I. Wolfing, Bernard P. Gee, Jason Porter, Alfredo Dubra, Ted H. Twietmeyer, Kamran Ahamd, Remy Tumbar, Fred Reinholz, and David R. Williams. In vivo fluorescence imaging of primate retinal ganglion cells and retinal pigment epithelial cells. *Opt. Express*, 14(16):7144–7158, Aug 2006.

[10] D. G. Bruns, D. G. Sandler, B. Martin, and G. Brusa. Design and prototype tests of an adaptive secondary mirror for the new 6.5-m single-mirror MMT. *Society of Photo-Optical Instrumentation Engineers (SPIE) Conference Series*, 2534:130–133, 08 1995.

[11] G. Brusa, A. Riccardi, S. Ragland, S. Esposito, C. del Vecchio, L. Fini, P. Stefanini, V. Biliotti, P. Ranfagni, P. Salinari, D. Gallieni, R. Biasi, P. Mantegazza, G. Sciocco, G. Noviello, and S. Invernizzi. Adaptive secondary P30 prototype: laboratory results. In D. Bonaccini and R. K. Tyson, editors, *Adaptive Optical System Technologies*, volume 3353 of *Society of Photo-Optical Instrumentation Engineers (SPIE) Conference Series*, pages 764–775, September 1998.

[12] A. Riccardi, G. Brusa, P. Salinari, D. Gallieni, R. Biasi, M. Andrighettoni, and H.M. Martin. Adaptive secondary mirrors for the large binocular telescope. *Adaptive Optical System Technologies II*, 4839(1):721–732, 2003.

[13] R. Arsenault, R. Biasi, D. Gallieni, A. Riccardi, P. Lazzarini, N. Hubin, E. Fedrigo, R. Donaldson, S. Oberti, S. Stroebele, R. Conzelmann, and M. Duchateau. A deformable secondary mirror for the VLT. In *Society of Photo-Optical Instrumentation Engineers (SPIE) Conference Series*, volume 6272 of *Society of Photo-Optical Instrumentation Engineers (SPIE) Conference Series*, July 2006.

[14] A. Riccardi, G. Brusa, P. Salinari, S. Busoni, O. Lardiere, P. Ranfagni, D. Gallieni, R. Biasi, M. Andrighettoni, S. Miller, and P. Mantegazza. Adaptive secondary mirrors for the Large binocular telescope. In R. K. Tyson & M. Lloyd-Hart, editor, *Society of Photo-Optical Instrumentation Engineers (SPIE) Conference Series*, volume 5169 of *Presented at the Society of Photo-Optical Instrumentation Engineers (SPIE) Conference*, pages 159–168, December 2003.

[15] Roberto Biasi, Mario Andrighettoni, Daniele Veronese, Valdemaro Biliotti, Luca Fini, Armando Riccardi, Paolo Mantegazza, and Daniele Gallieni. Lbt adaptive secondary electronics. In Peter L. Wizinowich and Domenico Bonaccini, editors, *SPIE*, volume 4839, pages 772–782, 2003.

[16] G. Brusa, A. Riccardi, P. Salinari, F. P. Wildi, M. Lloyd-Hart, H. M. Martin, R. Allen, D. Fisher, D. L. Miller, R. Biasi, D. Gallieni, and F. Zocchi. MMT adaptive secondary: performance evaluation and field testing. In *Adaptive Optical System Technologies II. Edited by Wizinowich, Peter L.; Bonaccini, Domenico*, volume 4839 of *Society of Photo-Optical Instrumentation Engineers (SPIE) Conference Series*, pages 691–702, February 2003.

[17] DW Miller and SCO Grocott. Robust control of the multiple mirror telescope adaptive secondary mirror. *Optical Engineering*, 38(8):1276 – 1287, 1999.

[18] C. R. Vogel and Q. Yang. Modeling and open-loop control of point-actuatied, continuous facesheet deformable mirrors. In *Society of Photo-Optical Instrumentation Engineers (SPIE) Conference Series*, volume 6272 of *Society of Photo-Optical Instrumentation Engineers (SPIE) Conference Series*, July 2006.

[19] Lucie Baudouin, Christophe Prieur, Fabien Guignard, and Denis Arzelier. Robust control of a bimorph mirror for adaptive optics systems. *Appl. Opt.*, 47(20):3637–3645, 2008.

[20] Thomas Ruppel, Michael Lloyd-Hart, Daniela Zanotti, and Oliver Sawodny. Modal trajectory generation for adaptive secondary mirrors in astronomical adaptive optics. In *Proc. IEEE International Conference on Automation Science and Engineering CASE 2007*, pages 430–435, 9 2007.

[21] T. Ruppel, W. Osten, and O. Sawodny. Model-based feedforward control of large deformable mirrors. *European Journal of Control*, 17(3):261–272, 2011.

[22] T. Ruppel, S. Dong, F. Rooms, W. Osten, and O. Sawodny. Feedforward control of deformable membrane mirrors for adaptive optics. *Transactions on Control Systems Technology*, Accepted for Publication, 2011.

[23] L. Meirovitch. *Analytical methods in vibration*. New York, NY.: The Mcmillan Company, 1967.

[24] Ronald P. Grosso and Martin Yellin. The membrane mirror as an adaptive optical element. *J. Opt. Soc. Am.*, 67(3):399–406, 03 1977.

[25] Luc Arnold. Optimized axial support topologies for thin telescope mirrors. *Optical Engineering*, 34(2):567–574, 1995.

[26] S. K. Ravensbergen, R. F. H. M. Hamelinck, P. C. J. N. Rosielle, and M. Steinbuch. Deformable mirrors: design fundamentals for force actuation of continuous facesheets. In Richard A. Carreras, Troy A. Rhoadarmer, and David C. Dayton, editors, *Proceedings of the SPIE*, volume 7466, page 74660G. SPIE, 2009.

[27] Frederick Bloom and Douglas Coffin. *Handbook of thin plate buckling and postbuckling.* Chapman & Hall/CRC, Boca Raton, FL, 2001.

[28] J Juillard and E Colinet. Modelling of nonlinear circular plates using modal analysis: simulation and model validation. *Journal of Micromechanics and Microengineering,* 16(2):448, 2006.

[29] M. Amabili. *Nonlinear vibrations and stability of shells and plates.* Cambridge University Press, Cambridge and New York, 2008.

[30] J.N. Reddy. *An introduction to the finite element method,* volume 2. McGraw-Hill New York, 1993.

[31] U.F. Meißner and A. Maurial. *Die Methode der finiten Elemente.* Springer Verlag, 2000.

[32] S. E. Winters, J. H. Chung, and S. A. Velinsky. Modeling and control of a deformable mirror. *Journal of Dynamic Systems, Measurement, and Control,* 124(2):297–302, 2002.

[33] C. del Vecchio. Supporting a magnetically levitated very thin meniscus for an adaptive secondary mirror: summary of finite-element analyses [3126-49]. In R. K. Tyson & R. Q. Fugate, editor, *Society of Photo-Optical Instrumentation Engineers (SPIE) Conference Series,* volume 3126 of *Presented at the Society of Photo-Optical Instrumentation Engineers (SPIE) Conference,* pages 397–+, October 1997.

[34] Sven Verpoort and Ulrich Wittrock. Actuator patterns for unimorph and bimorph deformable mirrors. *Appl. Opt.,* 49(31):G37–G46, 11 2010.

[35] Mauro Manetti, Marco Morandini, and Paolo Mantegazza. High precision massive shape control of magnetically levitated adaptive mirrors. *Control Engineering Practice,* 18(12):1386 – 1398, 2010.

[36] G. Agapito, S. Baldi, G. Battistelli, D. Mari, E. Mosca, and A. Riccardi. Automatic tuning of the internal position control of an adaptive secondary mirror. *European Journal of Control,* 17(3):273–289, 2011. Anglais.

[37] R.K. Tyson and B.W. Frazier. *Field guide to adaptive optics.* SPIE Press, 2004.

[38] K. Bush, D. German, B. Klemme, A. Marrs, and M. Schoen. Electrostatic membrane deformable mirror wavefront control systems: design and analysis. In J. D. Gonglewski, M. T. Gruneisen, & M. K. Giles, editor, *Society of Photo-Optical Instrumentation Engineers (SPIE) Conference Series,* volume 5553 of *Presented at the Society of Photo-Optical Instrumentation Engineers (SPIE) Conference,* pages 28–38, October 2004.

[39] E. Scott Claflin and Noah Bareket. Configuring an electrostatic membrane mirror by least-squares fitting with analytically derived influence functions. *J. Opt. Soc. Am. A,* 3(11):1833–1839, 1986.

[40] Arthur Menikoff. Actuator influence functions of active mirrors. *Appl. Opt.,* 30(7):833–838, 1991.

[41] K.F. Graff. *Wave motion in elastic solids*. Dover Pubns, 1991.

[42] E. Cerda and L. Mahadevan. Geometry and Physics of Wrinkling. *Physical Review Letters*, 90(7):074302–+, February 2003.

[43] V.Z. Vlasov.   *Allgemeine Schalentheorie und ihre Anwendung in der Technik*. Akademie-Verlag, 1958.

[44] R. Hamelinck, N. Rosielle, P. Kappelhof, B. Snijders, and M. Steinbuch. A large adaptive deformable membrane mirror with high actuator density.   *Society of Photo-Optical Instrumentation Engineers (SPIE) Conference Series*, 5490:1482–1492, October 2004.

[45] Steven Fortune and Gordon Wilfong. *Planning constrained motion*. ACM, New York, NY, USA, 1988.

[46] G. Brusa, A. Riccardi, V. Biliotti, C. del Vecchio, P. Salinari, P. Stefanini, P. Mantegazza, R. Biasi, M. Andrighettoni, C. Franchini, and D. Gallieni. Adaptive secondary mirror for the 6.5-m conversion of the Multiple Mirror Telescope: first laboratory testing results. In R. K. Tyson & R. Q. Fugate, editor, *Society of Photo-Optical Instrumentation Engineers (SPIE) Conference Series*, volume 3762 of *Presented at the Society of Photo-Optical Instrumentation Engineers (SPIE) Conference*, pages 38–49, September 1999.

[47] F. P. Wildi, G. Brusa, A. Riccardi, M. Lloyd-Hart, H. M. Martin, and L. M. Close. Towards first light of the 6.5m MMT adaptive optics system with deformable secondary mirror. In *Adaptive Optical System Technologies II. Edited by Wizinowich, Peter L.; Bonaccini, Domenico*, volume 4839 of *Proceedings of the SPIE*, pages 155–163, 02 2003.

[48] Gallieni, D., Tintori, M., Mantegazza, M., Anaclerio, E., Crimella, L., Acerboni, M., Biasi, R., Angerer, G., Andrigettoni, M., Merler, A., Veronese, D., Carel, J-L, Marque, G., Molinari, E., Tresoldi, D., Toso, G., Spanó, P., Riva, M., Mazzoleni, R., Riccardi, A., Mantegazza, P., Manetti, M., Morandini, M., Vernet, E., Hubin, N., Jochum, L., Madec, P., Dimmler, M., and Koch, F. Voice-coil technology for the e-elt m4 adaptive unit. In *1st AO4ELT conference - Adaptive Optics for Extremely Large Telescopes*, page 06002, 2010.

Chapter 5

Liquid Crystal Wavefront Correctors

Li Xuan, Zhaoliang Cao, Quanquan Mu, Lifa Hu and
Zenghui Peng

Additional information is available at the end of the chapter

1. Introduction

Liquid crystal (LC) was first discovered by the Austrian botanical physiologist Friedrich Reinitzer in 1888 [1]. It was a new state of matter beyond solid and liquid materials, having properties between those of a conventional liquid and those of a solid crystal. LC molecules usually have a stick shape. The average direction of molecular orientation is given by the director $\hat{n}$. When light propagates along the director $\hat{n}$, the refractive index is noted as the extraordinary index n_e, no matter the polarization direction (in the plane perpendicular to the long axis). However, the refractive index is different depending upon the polarization direction when light moves perpendicular to the director. When an electric field is employed, the LC molecule will be rotated so that the director $\hat{n}$ is parallel to the electric field. Due to the applied electric field, the LC molecular can be rotated from 0° to 90° and the effective refractive index is changed from n_e to n_o. As a result, the effective refractive index of LC can be controlled by controlling the strength of the electric field applied on the LC. The maximum change amplitude of the refractive index is birefringence $\Delta n = n_e - n_o$.

The properties discussed above allow LC to become a potential candidate for optical wavefront correction. A liquid crystal wavefront corrector (LCWFC) modulates the wavefront by the controllable effective refractive index, which is dependent on the electric field. As distinct from the traditional deformable mirrors, the LCWFC has the advantages of no mechanical motion, low cost, high spatial resolution, a short fabrication period, compactness and a low driving voltage. Therefore, many researchers have investigated LCWFCs to correct the distortions.

Initially, a piston-only correction method was used in LC adaptive optics (LC AOS) to correct the distortion. The maximum phase modulation equals Δn multiplying the thickness of the LC layer, and it is about 1μm. As reported [2], the pixel size was over 1mm and the number of pixels was about one hundred at that time. Because of the large pixel size, LCWFC not only loses the advantage of high spatial resolution but also mismatches the microlens array of the detector, which leads to additional spatial filtering in order to decrease the effect of the undetectable pixel for correction [3]. Moreover the small modulation amplitude makes it unavailable for many conditions. The thickness and Δn can be increased in order to increase the modulation amplitude. However, this will slow down the speed of the LCWFC.

Along with the development of LCWFC, an increasing number of commercial LC TVs are used directly for wavefront correction. Due to the high pixel density, the capacity for wavefront correction has been understood gradually by the researchers and the use of kinoform to increase the modulation amplitude is also possible [4-8]. A kinoform is a kind of early binary optical element which can be utilized in a high pixel density LCWFC. The wavefront distortion can be compressed into one wavelength with a 2π modulus of a large magnitude distortion wavefront. The modulated wavefront is quantified according to the pixel position of LCWFC. As discussed above, LCWFC only needs one wavelength intrinsic modulation amplitude to correct a highly distorted wavefront.

Many domestic and international researchers have devoted themselves to exploring LCWFCs from th 1970s onwards. In 1977, a LCWFC was used for beam shaping by I. N. Kompanets et al. [9]. S. T. Kowel et al. used a parallel alignment LC cell to fabricate a adaptive focal length plano-convex cylindrical lens in 1981 [10]. In 1984, he also realized a spherical lens by using two perpendicularly placed LC cells[11]. A LCWFC with 16 actuators was achieved in 1986 by A. A. Vasilev et al. and a one dimensional wavefront correction was realized [12]. Three years later, he realized beam adaptive shaping through 1296 actuators of an optical addressed LCWFC [13].

As a result, the LC AOS is becoming increasingly developed. In order to overcome the disadvantages of a traditional deformable mirror, such as a small number of actuators and high cost, D. Bonaccini et al. discussed the possibility of using LCWFC in a large aperture telescope [14, 15]. In 1995, D. Rensheng et al. used an Epson LC TV to perform a closed-loop adaptive correction experiment [16]. Although the twisted aligned LCWFC with the response time of 30ms was used, the feasibility of the LC AOS for wavefront correction was verified. Hence, many American [17-23], European [24-28] and Japanese [29] groups were devoted to the study of LC AOS. In 2002, the breakthrough for LC AOS was achieved and the International Space Station and various satellites were clearly observed [30]. In recent years, Prof. Xuan's group has completed series of valuable studies [31-42]. Recently, the applications of LCWFC have been extended to other fields, such as retina imaging [43-45], beam control [46-50], optical testing [41], optical tweezers [51-53], dynamic optical diffraction elements [54-57], tuneable photonic crystal fibre [58, 59], turbulence simulation [60, 61] and free space optical communications [62, 63].

The basic characteristics of a diffractive LCWFC are introduced in this chapter. The diffractive efficiency and the fitting error of the LCWFC are described first. For practical applications, the effects of tilt incidence and the chromatism on the LCWFC are expounded. Finally, the fast response liquid crystal material is demonstrated as obtaining a high correction speed.

2. Diffraction efficiency

2.1. Theory

A Fresnel phase lens model is used to approximately calculate the diffraction efficiency of the LCWFC. According to the rotational symmetry and periodicity along the r^2 direction, when the Fresnel phase lens is illuminated with a plane wave of unit amplitude, the complex amplitude of the light can be expressed as [64]:

$$f(r^2) = f(r^2 + jr_p^2) \tag{1}$$

where j is an integer and the period is r_p^2. Also, it can be expressed by the Fourier series:

$$f(r^2) = \sum_{n=-\infty}^{+\infty} A_n \exp[i2\pi nr^2 / r_p^2] \tag{2}$$

The distribution of the complex amplitude at the diffraction order n can be obtained [65]:

$$A_n = 1 / r_p^2 \int_0^{r_p^2} f(r^2) \exp[i2\pi nr^2 / r_p^2] dr^2 \tag{3}$$

For the Fresnel phase lens, the light is mainly concentrated on the first order ($n=1$). The diffraction efficiency of the Fresnel phase lens is defined as the intensity of the first order at its primary focus:

$$\eta = I(n = 1) = |A_1|^2 \tag{4}$$

If the phase distribution function $f(r^2)$ of the Fresnel phase lens can be achieved, the diffraction efficiency can be calculated by Eq. (3) and Eq. (4).

To correct the distorted wavefront, the 2π modulus should be performed first to wrap the phase distribution into one wavelength. Then, the modulated wavefront will be quantized. For a example, the wrapped phase distribution of a Fresnel phase lens is shown in Fig. 1(a). To a Fresnel phase lens, the 2π phase is always quantized with equal intervals. Assuming the height before quantization is h, the quantization level is N and the height of each quantized step is h/N. Figure 1(b) is a Fresnel phase lens quantized with 8 levels.

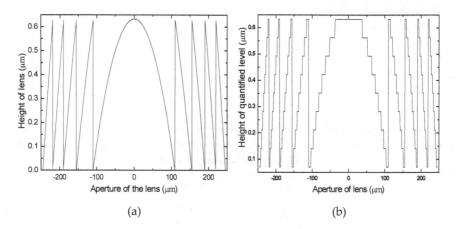

(a)   (b)

Figure 1. Phase distribution of a Fresnel phase lens: (a) 2π modulus; (b) quantized.

For a quantized Fresnel phase lens, the diffraction efficiency can be expressed as [66]:

$$\eta = \sin c^2 \left(1 / N\right) \tag{5}$$

Figure 2 shows the diffraction efficiency as a function of the quantization level for a Fresnel phase lens.

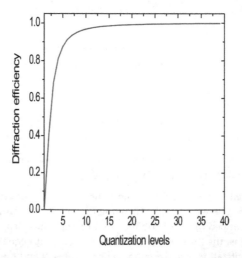

Figure 2. Diffraction efficiency as a function of the quantization level.

2.2. Effects of black matrix

A LCWFC always has a Black Matrix, which will cause a small interval between each pixel, as shown in Fig. 3. At the interval area, the liquid crystal molecule cannot be driven and then the phase modulation is different to the adjacent area. This will affect the diffraction efficiency of the LCWFC, as shown in Fig. 4. It is seen that the diffraction efficiency decreases by 6.4%, 8.8%, 9.5% and 9.7%, respectively for 4, 8, 16 and 32 levels, while the pixel interval is 1μm and the pixel pitch is 20μm. Consequently, the effect magnitude of the diffraction efficiency increases for a larger number quantified levels while the maximum decrease of the diffraction efficiency is about 10%.

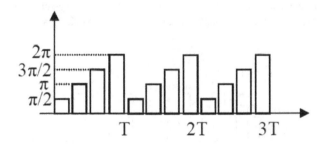

Figure 3. A Fresnel phase lens quantified by the pixel with a Black Matrix.

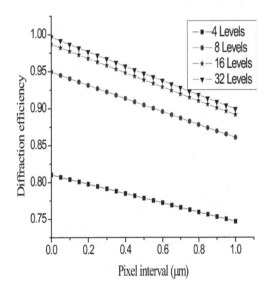

Figure 4. The diffraction efficiency as functions of the pixel interval for different quantified levels.

2.3. Mismatch between the pixel and the period

Because the pixel has a certain size P, the period T of a Fresnel phase lens cannot be divided exactly by the pixel, as shown in Fig. 5. This error is similar to the linewidth error caused by the lithography technique. For one period, the integer is n and the remainder is γ after T modulo P. If $\gamma \leq 0.5P$, there are n pixels in one period; on the contrary, there are $n+1$ pixels. As such, the maximum error is $0.5P$ for the first period. According to Eq. (3), the distribution of the complex amplitude of the first order can be acquired with the known phase distribution function in one period. Then, the diffraction efficiency can be obtained. As shown in Fig. 6, when the error of the first period changes from 0 to 0.5P, the diffraction efficiency decreases from 81% to 78.3%. The pixel number effect on the variation of the diffraction efficiency is also calculated while the error is 0.5P (Fig. 7). The decrease of the diffraction efficiency is 1% when the pixel number is 7. Accordingly, if the pixel number is not less than 7 in one period, the effect of the pixel size can be ignored.

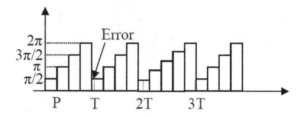

Figure 5. The mismatch between the pixel and the period of the Fresnel lens.

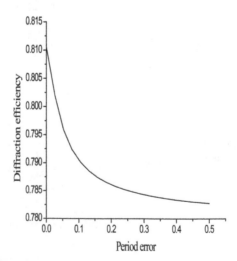

Figure 6. The diffraction efficiency as a function of the period error.

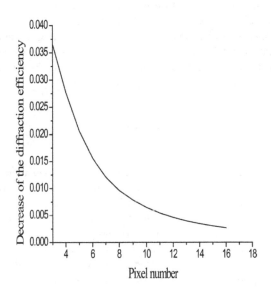

Figure 7. The decrease of the diffraction efficiency as a function of the pixel number.

2.3.1. Wavefront compensation error

The wavefront compensation error always exists due to the finite number of the wavefront correction element used for the correction of the atmospheric turbulence. Hudgin gave the relationship between the compensation error and the actuator size as follows [67]:

$$f = \alpha(\frac{r_s}{r_0})^{5/3} \tag{6}$$

where r_s is the actuator spacing, r_0 is the atmosphere coherence length and α is a constant depending on the response function of the actuator. For continuous surface deformable mirrors (DMs), the response function of the actuator is a Gaussian function and α ranges from 0.3-0.4 [68]. For a piston-only response function, α is 1.26 [69]. Researchers always use a piston-only response function to evaluate a LCWFC and have proved that the actuators need to be 4-5 times as large as that of the DM's [69, 70]. However, the case is totally different when a diffractive LCWFC (DLCWFC) is used where the kinoform or phase wrapping technique is employed to expand the correction capability [71, 72]. Therefore, Eq. (6) is not suitable any more.

2.4. The effect of quantization on the wavefront error

Firstly, the wavefront error generated during the phase wrapping due to quantization is considered. Since a LCWFC is a two-dimensional device, the quantification is performed along the x and y axes by taking the pixel as the unit. According to the diffraction theory, the correction precision as a function of the quantization level can be deduced [73]. If the pixel size is not considered, the root mean square (RMS) error of the diffracted wavefront as a function of the quantization level can be simplified as [73]:

$$\Delta W = \frac{\lambda}{2\sqrt{3}N} \tag{7}$$

Where N is the quantization level and λ is the wavelength. If $N=30$, then the RMS error can be as small as $\lambda/100$. For $N=8$, RMS=0.036λ and the corresponding Strehl ratio is 0.95. Figure 8 shows the diffracted wavefront RMS error as a function of the quantization level N. As can be seen, the wavefront RMS error reduces drastically at first, and then approaches to a constant gradually when the quantization level becomes greater than 10. The wavefront RMS error can be calculated for a known quantization level on a wavefront. For a DLCWFC, the wavefront compensation error is directly determined by the quantization level without any need to consider the pixel number. Therefore, the distribution of the quantization level on the atmospheric turbulence should be calculated first for a given pixel number, telescope aperture and atmospheric coherence length, and then the wavefront compensation error can be calculated by using Eq. (7).

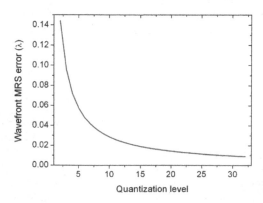

Figure 8. The wavefront RMS error as a function of the quantization level.

2.5. Zernike polynomials for atmospheric turbulence

Kolmogorov turbulence theory is employed to analyse the distribution of the quantization level across an atmospheric turbulence wavefront. Noll described Kolmogorov turbulence by using Zernike polynomials [74]. According to him, Zernike polynomials are redefined as:

$$
\begin{aligned}
Z_{even\ j} &= \sqrt{2(n+1)}R_n^m(\rho)\cos(m\theta),\ m \neq 0 \\
Z_{odd\ j} &= \sqrt{2(n+1)}R_n^m(\rho)\sin(m\theta),\ m \neq 0 \\
Z_j &= \sqrt{(n+1)}R_n^0(\rho),\qquad m=0
\end{aligned}
\tag{8}
$$

Where:

$$
R_n^m(\rho) = \sum_{s=0}^{(n-m)/2} \frac{(-1)^s(n-s)!}{s!\left[(n+m)/2 - s\right]!\left[(n-m)/2 - s\right]!} \cdot r^{n-2s}
\tag{9}
$$

The parameters n and m are integral and have the relationship $m \leq n$ and $n - |m| = even$. An atmospheric turbulence wavefront can be described by using a Kolmogorov phase structure function, as below [74]:

$$
D(r) = 6.88\left(\frac{r}{r_0}\right)^{5/3}
\tag{10}
$$

By combining the phase structure function and the Zernike polynomials, the covariance between the Zernike polynomials Z_j and $Z_{j'}$ with amplitudes a_j and $a_{j'}$ can be deduced as [75]:

$$
\langle a_j a_{j'} \rangle =
\begin{cases}
\dfrac{K_{zz} \cdot \delta_{mm'} \Gamma\left[(n+n'-5/3)/2\right](D/r_0)^{5/3}}{\Gamma\left[(n-n'+17/3)/2\right]\Gamma\left[(n'-n+17/3)/2\right]\Gamma\left[(n+n'+23/3)/2\right]} & j-j' = even \\
0, & j-j' = odd
\end{cases}
\tag{11}
$$

where $K_{zz'} = 2.698(-1)^{(n+n'-2m)/2}\sqrt{(n+1)(n'+1)}$ and D is the telescope diameter. $\delta_{mm'}$ is the Kronecker delta function. By using Eq. (11), the coefficients of the Zernike polynomials can be easily computed. If the first J modes of the Zernike polynomials are selected, the atmospheric turbulence wavefront is represented as:

$$
\phi_t = \sum_{j=1}^{J} a_j Z_j
\tag{12}
$$

Therefore, the atmospheric turbulence wavefront Φ_t can be calculated by using Eqs. (11) and (12). As the phase wrapping technique is employed, the atmospheric turbulence wavefront can be wrapped into 2π and quantized and thus the distribution of the quantization level across a telescope aperture D can be determined.

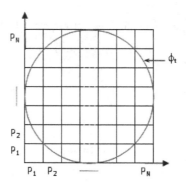

Figure 9. The field of the DLCWFC - the circle represents the wavefront of atmospheric turbulence and $P_1...P_N$ are the pixel numbers of the DLCWFC.

2.6. Calculation of the required pixel number of DLCWFCs

In practice, people hope to calculate the desired pixel number of a DLCWFC expediently for a given telescope aperture D, a quantization level N, and an atmospheric coherence length r_0. Therefore, it is necessary to deduce the relation between the pixel number of the DLCWFC and D, N and r_0. As shown in Fig. 9, the DLCWFC aperture can be represented by the pixel number across the aperture, which is called P_N. The circle represents the atmospheric turbulence wavefront Φ_t. Since the atmospheric turbulence wavefront is random, the ensemble average $<\Phi_t>$ should be used in the calculation. The modulated and quantized atmospheric turbulence wavefront can be expressed as:

$$\text{mod}(\langle \phi_t \rangle) = f(N, D, r_0, \langle P_N \rangle) \tag{13}$$

where mod( ) denotes the modulo 2π. If $<\Phi_t>$ is known, $<P_N>$ can be expressed as a function of the telescope aperture D, the quantization level N, and the atmospheric coherence length r_0. By using Eqs. (11) and (12), $<\Phi_t>$ can be calculated and the first 136 modes of the Zernike polynomials are used in the calculation. For the randomness of the atmospheric turbulence wavefront, different quantization levels are used during the quantization, depending upon the fluctuation degree of the wavefront. Here, N is defined as the minimum quantization level so that the sum of those quantization levels greater than N should occupy 95% of the quantization levels included in the atmospheric turbulence wavefront. Fifty atmospheric turbulence wavefronts are used to achieve the statistical results. First, the relation between the pixel number P_N and the telescope aperture D is calculated for r_0=10cm and N=16, as shown in Fig. 10. It can be seen that $<P_N>$ is a linear function of D when N and r_0 are fixed. That is to say, the larger the aperture of the telescope, the more the pixel number will be needed if a DLCWFC is used to correct the atmospheric turbulence. Specifically, for a telescope with a diameter of 2 metres, the total pixel number will be 96×96, while for a telescope with a diameter of 4 metres the total pixel number will be 168×168. P_N as a function of N is also computed for r_0=10cm and D =2 m, as shown in Fig.11. It illustrates that when D and r_0 are fixed, $<P_N>$ is a linear function of N. This means that the more that the quantization level is used, the more the pixel number will be needed.

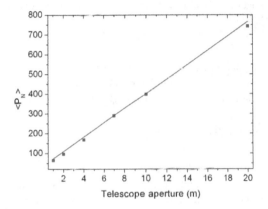

Figure 10. as a function of the telescope aperture D - ■ represents the calculated data for r_0=10 cm and N=16, and the solid curve represents the fitted data.

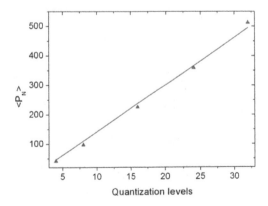

Figure 11. as a function of the quantization level N - ▲ represents the calculated data for D=2 m and r_0=10 cm, and the solid curve represents the fitted data.

The relationship between <P_N> and r_0 is also calculated with the variables N and D. Figure 12 shows only three curves with three pairs of fixed N and D. This time, the relationship is not a linear function anymore but an exponential function. With more pairs of N and D fixed, more curves can be obtained, but these are not shown in the figure. The relationship between <P_N> and r_0 can be expressed by the following formula:

$$\langle P_N \rangle = A + B r_0^{-6/5} \tag{14}$$

where A and B are the coefficients. A is only related to N and can be expressed as $A=6.25N$. As <P_N> is a linear function of D and N, the coefficient B can be expressed as:

$$B = a + bN + cD + dND \tag{15}$$

where a, b, c and d are the coefficients. By substituting the known value of N, D and the calculated coefficient B, the value of a, b, c and d is determined to be 15, -23, -150 and 91, respectively, by using the least-square method. Thus, <P_N> can be expressed as:

$$\langle P_N \rangle = 6.25N + \left(15 - 1.5D - 23N + 0.91ND\right) r_0^{-6/5} \tag{16}$$

where the units of D and r_0 is centimetres. The total pixel number of the DLCWFC can be calculated by using $P_N \times P_N$. By combining Eqs. (7) and (16), the compensation error of the DLCWFC can be evaluated for the atmospheric turbulence correction. These two formulas are not suitable for modal types of LCWFCs [76] or other types that do not use the diffraction method to correct the atmospheric turbulence.

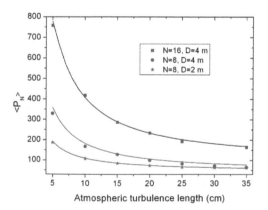

Figure 12. as a function of the atmosphere coherence length r_0 - the line is the fitted curve and ■, ● and ✳ represent the computed data with $N=16$ and $D=4$ m, $N=8$ and $D=4$ m, and $N=8$ and $D=2$ m, respectively.

Normally, the quantization level of 8 is suitable for the atmospheric turbulence correction for three reasons. Firstly, a higher correction accuracy can be obtained. When $N=8$, the RMS error can be reduced down to 0.035λ and the Strehl ratio can be increased to 95%. Secondly, a higher diffraction efficiency can be obtained. According to the diffractive optics theory [73], the diffraction efficiency is as large as 95% for $N=8$. Finally, the total pixel number can be controlled within a reasonable range. Of course, a smaller wavefront RMS error and a higher diffraction efficiency can be achieved with a larger quantization level. But, in that case, the required pixel number of the DLCWFC will be increased drastically, which will lead to a significantly slower computation and data transformation rate of the LCAOS. Fig. 13 shows the relation between P_N, D and r_0 for $N=8$. As can be seen, the desired pixel number apparently increases when the atmospheric coherence length becomes smaller and the telescope aperture becomes larger. For instance, the total pixel number of the DLCWFC is 1700×1700 when $r_0=5$ cm and $D=20$ m. However, if $r_0=10$ cm, the total pixel number can be reduced down to 768×768. Therefore, the strength of the atmosphere turbulence is a key factor which must be considered when designing the LCAOS for a ground-based telescope.

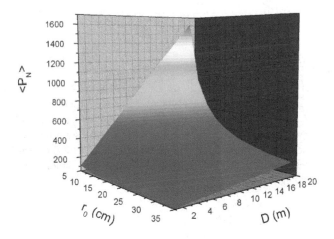

Figure 13. as functions of the atmosphere coherence length r_0 and the telescope aperture D for $N=8$.

2.6.1. Tilt incidence

Currently, reflective LCWFC devices [77-79], such as liquid crystal on silicon (LCOS) devices, are especially attractive because of their small fill factor, high reflectivity and short response time. To separate the incident beam from the reflected beam for a reflective LCWFC, the incident light should go to the LCWFC with a tilt angle. Alternatively, the incident light is perpendicular to the LCWFC and a beam splitter is placed before the LCWFC to separate the reflected and incident beams. However, the second method will result in a 50% loss in each direction, reducing output power to 25% of the input. To avoid the energy loss, the tilt incidence is a suitable method for a LCWFC. However, the tilt incidence will affect the phase modulation and the diffraction efficiency of the LCWFC. A reflective LCWFC model is selected to perform the analysis and the acquired results are suitable for the transmitted LCWFC.

2.7. Effect of the tilt incidence on the phase modulation of the LCWFC

In order to simplify the model of the reflective LCWFC, the border effect is neglected and all of the molecules have the same tilt angle. The simplified model is shown in Fig.14. The former board is glass and the back is silicon. The liquid crystal molecule is aligned parallel to the board. The tilt incident angle is θ'. The liquid crystal material is a uniaxial birefringence material - it has an ordinary index n_o and the extraordinary index $n_e(\theta)$. $n_e(\theta)$ can be obtained with the index ellipsoid equation [41]:

$$n_e(\theta) = \frac{n_o n_e}{\left(n_o^2 \cos^2 \theta + n_e^2 \sin^2 \theta\right)^{1/2}}, \tag{17}$$

where θ represents the tilt angle of the molecule and n_e is the off-state extraordinary refractive index. Assume that the LCWFC without the applied voltages and the polarization direction is the same as the LC director. For the tilt incidence as shown in Fig.14, it is equivalent to the rotation of the LC director with an angle θ'. Hence, although the tilt angle of the molecule is zero, the extraordinary refractive index is changed to $n_e(\theta')$ with the tilt incidence. Furthermore, the tilt incidence will change the transmission distance of the light in the liquid crystal cell with a factor of $1/\cos\theta'$. Consequently, the phase modulation with the tilt incidence and no applied voltages can be expressed as:

$$P_{tilt} = \frac{2\pi(n_e(\theta') - n_o)d}{\lambda \cos \theta'}. \tag{18}$$

If the pre-tilt angle of the liquid crystal molecule is considered, Eq.(18) can be rewritten as:

$$P_{tilt} = \frac{2\pi(n_e(\theta' + \theta_0) - n_o)d}{\lambda \cos \theta'}, \tag{19}$$

where θ_0 is the pre-tilt angle, d is the thickness of the liquid crystal cell and λ is the relevant wavelength.

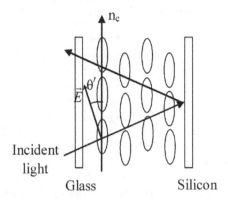

Figure 14. Simplified model of the reflective LCWFC with tilt incidence.

For n_e=1.714, n_o=1.516, λ=633nm and d=1.6µm, the phase modulation as a function of the incident angle is shown in Fig.15. The simulated results show that the phase modulation is reduced by at most 1% for incident angles under 6°. The measured result is also shown in the figure. The trends of the simulated and measured curves are similar. The difference of in phase shift might be caused by the border effect. For the actual liquid crystal cell, a rubbing polyimide (PI) film is used to align the liquid crystal molecules. The PI layer will anchor the liquid crystal molecules at the border; this causes the tilt angle of the liquid crystal molecule at the interface to be different from the centre. The simulated and measured results indicate that the LCWFC may be used with a small tilt angle.

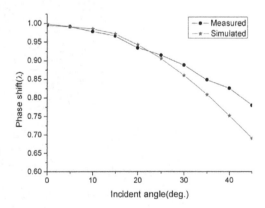

Figure 15. Phase shift as a function of the incident angle - -●- is the measured curve and -*- represents the simulated data.

2.8. The effect of pixel crossover on the phase modulation

For the tilt incidence, shown in Fig.16, the incident light in one pixel could transmit through an adjacent pixel, which is called pixel crossover. The maximum error of the pixel crossover is W. The pixel crossover will also affect the phase modulation of the LCWFC. Because each pixel is an actuator with a corresponding phase modulation, the light should go through just one pixel so as to control the phase modulation accurately. For a 19μm pixel size and d=1.6μm, W as a function of the incident angle is shown in Fig.17. The results show that W=0.33μm for a tilt incident angle of 6°. For a pixel with a size of P, the ratio of the light which transmits through adjacent pixels can be expressed with W/P. If the ratio equals to zero, it illustrates that the light is the vertical incidence and that it goes through just one pixel. For an incident angle of 6°, the ratio is only 1.77% and it may be ignored. As such, the LCWFC may be used at the tilt incidence condition with a little tilt angle.

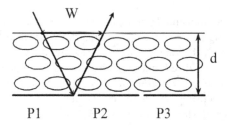

Figure 16. Illustration of the pixel crossover - P1, P2 and P2 are pixels, d is the thickness of the cell.

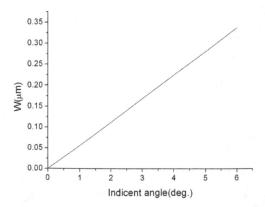

Figure 17. The pixel crossover W as a function of the incident angle.

2.9. Diffraction efficiency with tilt incidence

Because the phase of each pixel changes with the tilt incidence, the diffraction efficiency will decrease [64]. The Fresnel phase lens model [71] is used to calculate the change of the diffraction efficiency and 16 quantified levels are selected. The simulated results show that at an incident angle of 6°, the diffraction efficiency is reduced by 3% (Fig.18). For the incident angles less than 3°, the reduction in diffraction efficiency is less than 1% - a negligible loss for most applications.

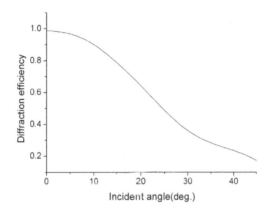

Figure 18. Diffraction efficiency as a function of the incident angle.

2.9.1. Chromatism

The chromatism of the LCWFC includes refractive index chromatism and quantization chromatism. Refractive index chromatism is caused by the LC material, and is generally called dispersion. Meanwhile, quantization chromatism is caused by the modulo 2π of the phase wrapping. Theoretically, the LCWFC is only suitable for use in wavefront correction for a single wavelength and not on a waveband due to chromatism. However, if a minor error is allowed, LCWFC can be used to correct distortion within a narrow spectral range.

2.10. Effects of chromatism on the diffraction efficiency of LCWFC

The measured birefringence dispersion of a nematic LC material (RDP-92975, DIC) is shown in Fig. 19. It can be seen that the birefringence Δn is dependent on the wavelength and the dispersion of the LC material is particularly severe when the wavelength is less than 500 nm. Since a phase wrapping technique is used, the phase distribution should be modulo 2π, and it should then be quantized [71]. Assuming that the quantization wavelength is λ_0, the thickness of the LC layer is d, and V_{max} denotes the voltage needed to obtain a 2π phase modulation, such that the maximum phase modulation of the LCWFC can be expressed as:

$$\Delta\varphi_{max}(\lambda_0) = 2\pi \frac{\Delta n(\lambda_0, V_{max})d}{\lambda_0} = 2\pi \qquad (20)$$

For any other wavelength λ, it can be rewritten as:

$$\Delta\varphi_{max}(\lambda) = 2\pi \frac{\Delta n(\lambda, V_{max})d}{\lambda} \qquad (21)$$

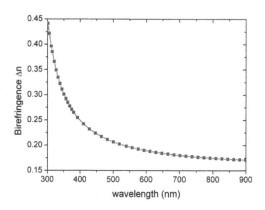

Figure 19. The birefringence Δn as a function of the wavelength.

For a quantization wavelength of 550 nm, 633 nm and 750 nm, the variation of the maximum phase modulation as a function of wavelength is shown in Fig. 20. Assuming that the deviation of the phase modulation is 0.1, for λ_0=550, 633 and 750 nm, the corresponding spectral ranges are calculated as 520–590 nm, 590–690 nm and 690–810 nm, respectively. If a 10% phase modulation error is acceptable, then the LCWFC can only be used to correct the distortion for a finite spectral range.

The variation of Δn and λ affects the diffraction efficiency of the LCWFC. Using the Fresnel phase lens model, the diffraction efficiency for any other wavelength λ can be described as [80]:

$$\eta = \left| \frac{\sin(\pi d\Delta n(\lambda, V_{max})/\lambda)}{\pi(d\Delta n(\lambda, V_{max})/\lambda - 1)} \right|^2 \qquad (22)$$

The effects of Δn and λ on the diffraction efficiency are shown in Fig. 21. For λ_0 = 550 nm, 633 nm and 750 nm, and their respective corresponding wavebands of 520–590 nm, 590–690 nm and 690–810 nm, the maximum energy loss is 3%, which is acceptable for the LC AOS. Although only one kind of LC material is measured and analysed, the results are helpful in the use of LCWFCs because almost all the nematic LC materials have similar dispersion characteristics.

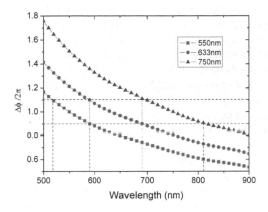

Figure 20. The phase modulation as a function of the wavelength for $\lambda_0 = 550$ nm, 633 nm and 750 nm, respectively - the two horizontal dashed lines indicate the phase deviation range while the four vertical dashed lines illustrate three sub-wavebands of 520–590 nm, 590–690 nm and 690–820 nm, respectively.

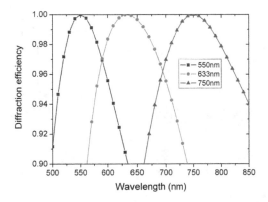

Figure 21. The diffraction efficiency as a function of wavelength for $\lambda_0 = 550$ nm, 633 nm and 750 nm, respectively.

2.11. Broadband correction with multi-LCWFCs

The above calculated results show that it is only possible to correct the distortion in a narrow waveband using only one LCWFC. Therefore, to realize the distortion correction in a broadband - such as 520–810 nm - multi-LCWFCs are necessary; each LCWFC is responsible for the correction of different wavebands and then the corrected beams are combined to realize the correction in the whole waveband. The proposed optical set-up is shown in Fig. 22, where a polarized beam splitter (PBS) is used to split the unpolarized light into two linear polarized beams. An unpolarized light can be looked upon as two beams with cross polarized states. Because the LCWFC can only correct linear polarized light, an unpolarized incident light can only be corrected in one polarization direction while the other polarized beam will not be corrected. Therefore, if a PBS is placed following the LCWFC, the light will be split into two linear polarized beams: one corrected beam goes to a camera; the other uncorrected beam is used to measure the distorted wavefront by using a wavefront sensor (WFS). This optical set-up looks like a closed loop AOS, but it is actually an open-loop optical layout. This LC adaptive optics system must be controlled through the open-loop method [31, 81]. Three dichroic beam splitters (DBSs) are used to acquire different wavebands. A 520–810 nm waveband is acquired by using a band-pass filter (DBS1). This broadband beam is then divided into two beams by a long-wave pass filter (DBS2). Since DBS2 has a cut-off point of 590 nm, the reflected and transmitted beams of the DBS2 have wavebands of 520–590 nm and 590–820 nm, respectively. The transmitted beam is then split once more by another long-wave pass filter (DBS3) whose cut-off point is 690 nm. Through DBS3, the reflected and transmitted beams acquire wavebands of 590–690 nm and 690–810 nm, respectively. Thus, the broadband beam of 520–810 nm is divided into three sub-wavebands, each of which can be corrected by an LCWFC. After the correction, three beams are reflected back and received by a camera as a combined beam. Using this method, the light with a waveband of 520–810 nm can be corrected in the whole spectral range with multi-LCWFCs.

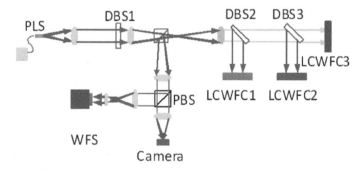

Figure 22. Optical set-up for a broadband correction - PLS represents a point light source, PBS is a polarized beam splitter, DBS means dichroic beam splitter, DBS1 is a band-pass filter, and DBS2 and DBS3 are long-pass filters.

The broadband correction experimental results are shown in Fig. 23. A US Air Force (USAF) resolution target is utilized to evaluate the correction effects in a broad waveband. Firstly, the waveband of 520–590 nm is selected to perform the adaptive correction. After the correction, the second element of the fifth group of the USAF target is resolved, with a resolution of 27.9 μm (Fig.23(b)). Considering that the entrance pupil of the optical set-up is 7.7 mm, the diffraction-limited resolution is 26.4 μm for a wavelength of 550 nm. Thus, a near diffraction-limited resolution has been achieved. Figure 23(c) shows the resolving ability for a waveband of 590–690 nm. The first element of the fifth group is resolved and the resolution is 31.25 μm, which is near the diffraction-limited resolution of 30.4 μm for a 633 nm wavelength. The corrected result for 520–690 nm is shown in Fig. 23(d). The first element of the fifth group can also be resolved. These results show that a near diffraction-limited resolution of an optical system can be obtained by using multi-LCWFCs.

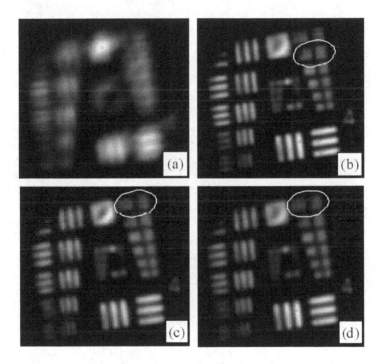

Figure 23. Images of the resolution target for different wavebands: (a) no correction; (b) 520–590 nm; (c) 590–690 nm; (d) 520–690 nm - the circular area represent the resolving limitation.

2.11.1. Fast response liquid crystal material

In applications of LCWFCs, the response speed is a key parameter. A slow response will significantly decrease the bandwidth of LC AOS. To improve the response speed, dual-frequency and ferroelectric LCs have been utilized to fabricate the LCWFC [82, 83]. However, there are some shortcomings with these fast materials. The driving voltage of the dual-frequency LCWFC is high and it is incompatible with the very large scale integrated circuit. The phase modulation of the ferroelectric LCWFC is very slight and it is hard to correct the distortions. Nematic LCs have no such problems. However, its response speed is slow. In this section, we introduce how to improve the response speed of nematic LCs.

For a nematic LC device, the response time of LC can be described by the following equations when the LC cell is in parallel-aligned mode [84]:

$$\tau_{rise} = \frac{\gamma_1 d^2}{K_{11}\pi^2 \ \left(V/V_{th}\right)^2 - 1} \tag{23}$$

$$\tau_{decay} = \frac{\gamma_1 d^2}{K_{11}\pi^2} \tag{24}$$

where γ_1 is the rotational viscosity, V and V_{th} are turn-on driving and threshold voltage, K_{11} is the elastic constant and d is the thickness of the LC cell. Generally, the rise time can be decreased by the overdriving method. However the decay time particularly depends upon the intrinsic parameters of LC devices, which are the key factors for response improvement. From Eq. (24), the smaller visco-elastic coefficient (γ_1/K_{11}) and d is, the shorter the response time is. However, it is necessary to keep the phase retardation ($d \times \Delta n$) to exceed (or equal) one wavelength for a LCWFC, and then the cell gap can only been reduced to a limited value for a constant birefringence (Δn). The higher birefringence of LC materials enables a thinner cell gap to be used while keeping the same phase retardation and improves the response performance of the LCWFC. Therefore, the LC materials with high Δn and low γ_1/K_{11} are required to have a fast response.

In the study of fast response LC materials, a concept of 'figure-of-merit' (FoM) is adopted to evaluate different LC compounds [85], as shown as Eq. (25). A LC material with a high FoM value will provide a short response time:

$$FoM = K_{11}\Delta n^2 / \gamma_1 \tag{25}$$

2.12. Nematic liquid crystal molecular design

In practice, some simple empirical rules together with a trial are usually used to help with the molecular design and mixing, such as LC compounds with a tolane and biphenyl group with a large Δn and a moderate γ_1. Recently, some computer simulation-based theoretical studies have been performed in order to shed light on the connections between macroscopic properties and molecular structure. A notable advantage of simulation is to predict the properties of a nematic LC material with optimal molecular configurations instead of costly and time-consuming experimental synthesis. In the study of fast response LCs, theoretical methods are used to analyse the rotational viscosity and Δn of a specific chemical structure.

In the study of the rotational viscosity (RVC) of nematic liquid crystals, Zhang et al. [86] adopt two statistical-mechanical approaches proposed by Nemtsov-Zakharov (NZ) [87] and Fialkowski (F) [88]. The NZ approach is based on the random walk theory. It is a correction of its predecessor in considering the additional correlation of the stress tensors with the director and the fluxes with the order parameter tensor, except for the autocorrelation of the microscopic stress tensor.

In Fig.24, the RVC of the nematic liquid crystal E7 is shown as a function of temperature. The experimental rotational viscosity decreases with temperature, and similar variations from NZ and F's theoretical methods are also obtained. The calculated NZ and F rotational viscosities are in the same order of magnitude as the experimental values. The larger the number of molecules, the longer the simulation time, and the revised force field for liquid crystals is expected to be helpful in improving this prediction.

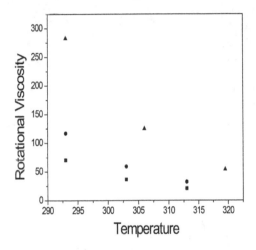

Figure 24. Temperature dependence of the rotational viscosity for E7, ■, the NZ method, ●, the F method, ▲, and the experiment.

The birefringence and dielectric anisotropy can be calculated by the Vuks equation and the Maier-Meier theory, respectively, and these calculated values have a good correlation with the experimental data in Ref. 89. In all, these approaches comprise a unique molecular design method for fast response LCs.

2.13. Chemosynthesis of fast response LC materials

In order to achieve fast LC material, researchers have synthesized a series of high birefringence LC materials with a linear shape and a long conjugated group. Gauza et al. first synthesized and reported a biphenyl, cyclohexyl- biphenyl isothiocyanato (NCS) LC material in which Δn is 0.2-0.4 and the rotational viscosity is about 10 ms μm^{-2}. The chemical structures are shown in Fig. 25. Moreover, they perform a comparison with a commercial E7 mixture. At 70°C, the *FoM* of the NCS mixture has a factor ten higher than that of E7 at 48°C [90].

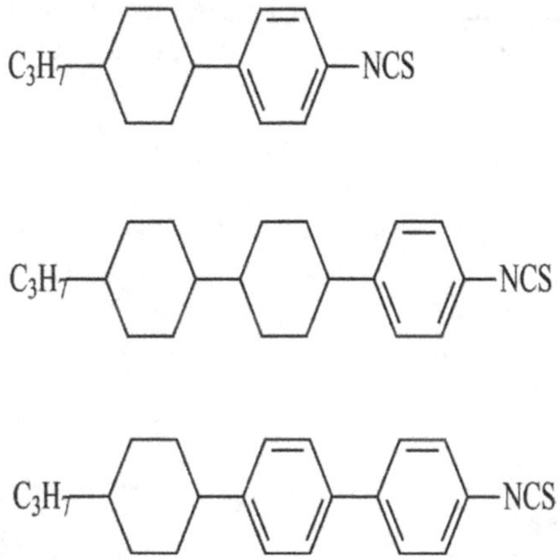

Figure 25. Chemical structures of biphenyl, cyclohexyl- biphenyl isothiocyanato LC materials.

In 2006, Gauza [91] provided one type of NCS LC material with unsaturated groups. The LC chemical structures are shown in Fig. 26: the final two NCS LC mixtures show a Δn value of 0.25 and 0.35; a viscosity factor of about 6 ms μm^{-2}; FoM values of 10.1 and 18.7 $\mu m^2 s^{-1}$. The response speed of such a LC material can be as low as 640 μs with a LC thickness of 2 μm at 35°C.

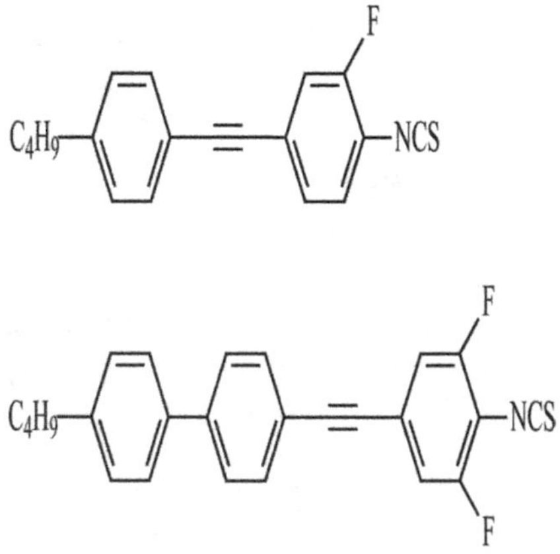

Figure 26. Chemical structures of NCS LC materials with unsaturated groups.

The high birefringence isothiocyanato LC with a tolane or terphenyl group can usually be synthesized via a couple reaction; the chemical reaction route was shown in Fig. 27 [92]:

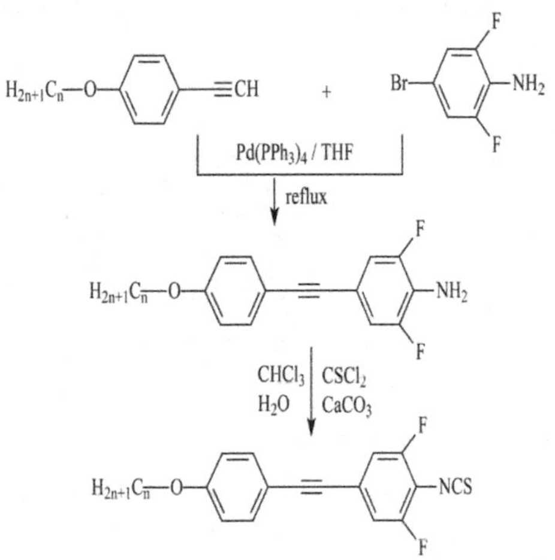

Figure 27. The synthesis of isothiocyanato compounds using Suzuki coupling.

In Gauza et al., in subsequent research, a series of fluro-substituted NCS LC materials with a Δn up to 0.5 at room temperature was developed, and some of them show better response performance [93], the chemical structures are shown in Fig. 28:

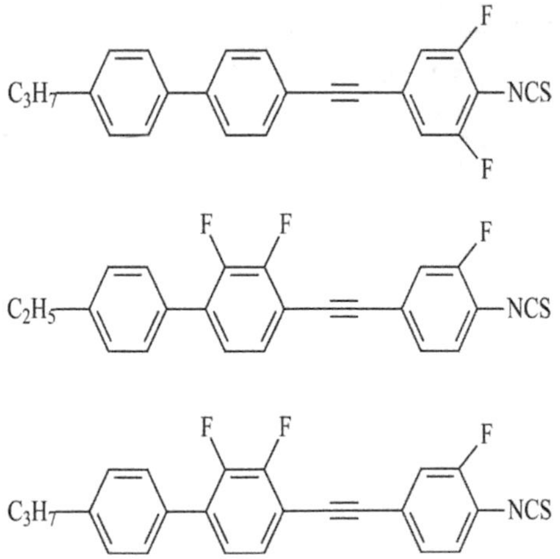

Figure 28. Chemical structures of NCS LC materials with high birefringence.

In the research of isothiocyanate tolane LC compounds, Peng et al. prepared a NCS LC compound via an electronation reaction. The reaction route is shown in Fig. 29. Compared to the conventional couple reaction method, this synthesis route improves the total reaction yield [94].

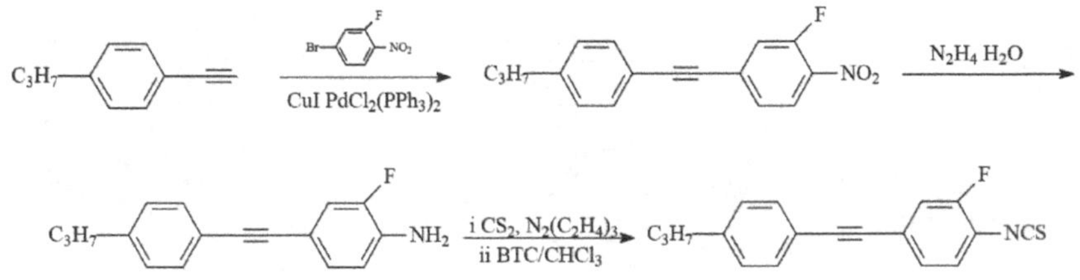

Figure 29. The synthesis of isothiocyanato tolane LC compound using electronation reaction.

It has rarely been reported that LCs with a very low rotational viscosity were mixed to high Δn LCs in order to improve response performance. However, Peng et al. introduce a type of difluorooxymethylene-bridged (CF_2O) LCs with a very low rotational viscosity so as to improve the response performance of NCS LCs. The chemical structure is shown in Fig. 30. When the material was mixed to NCS LCs with a high Δn, the visco-elastic coefficient of mixture decreased noticeably, the LC mixture approximately maintained high birefringence, and the *FoM* value increased from 14.8 to 16.9 $\mu m^2 s^{-1}$ at 7% concentration [95].

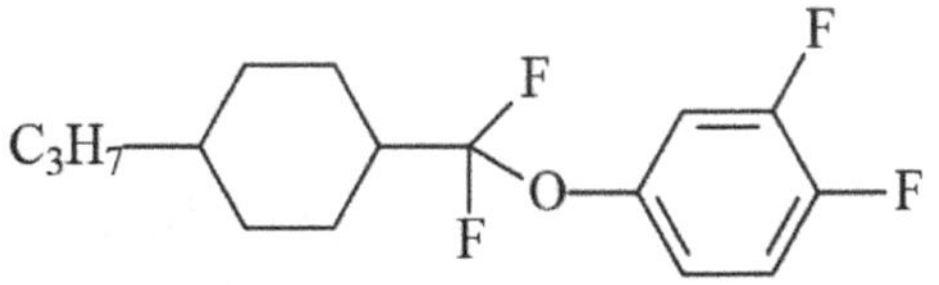

Figure 30. Chemical structure of difluorooxymethylene- bridged LC compound.

Acknowledgements

This work is supported by the National Natural Science Foundation of China, with Grant Nos. 50703039, 60736042, 11174274 and 11174279.

Author details

Li Xuan, Zhaoliang Cao, Quanquan Mu, Lifa Hu and Zenghui Peng

State Key Laboratory of Applied Optics, Changchun Institute of Optics, Fine Mechanics and Physics, Chinese Academy of Sciences, Jilin Changchun, China

References

[1] F. Reinitzer, Monatsh. Chemie, 9, 421–425 (1888).

[2] D. Dayton, S. Browne, J. Gonglewski. SPIE, 5894, 58940M.1-58940M.6 (2005).

[3] D. C. Dayton, S. L. Browne, J. D. Gonglewski, S. R. Restaino. Appl. Opt., 40, 2345-2355 (2001).

[4] J. Amako, T. Sonehara. Applied Optics, 30, 4622-4628 (1991).

[5] M. T. Gruneisen, L. F. DeSandre, J. R. Rotge, R. C. Dymale, D. L. Lubin. Opt. Eng., 43, 1387-1393 (2004).

[6] M. T. Gruneisen, R. C. Dymale, M. B. Garvin. SPIE, 5894, 589412.1-589412.10 (2005).

[7] M. T. Gruneisen, L. F. Desandre, J. R. Rotge, R. C. Dymale, D. L. Lubin. Opt. Eng., 43, 1387-1393 (2004).

[8] M. T. Gruneisen, R. C. Dymale, J. R. Rotge, L. F. Desandre, D. L. Lubin. Opt. Eng., 44, 023201.1-023201.9 (2005).

[9] I. N. Kompanets, Zarubezh. Radioelektron, 4, 46 (1977).

[10] S. T. Kowel, D. S. Cleverly. "A Liquid Crystal Adaptive Lens," in Proceedings, NASA Conference on Optical Information Processing for Aerospace Applications, Hampton, Va. (1981).

[11] S. T. Kowel, P. Kornreich, A. Nouhi. "Adaptive spherical lens," Applied Optics, Vol. 23, No. 16, 2774-2777 (1984).

[12] A. A. Vasil'ev, A. F. Naumov, V. I. Shmal'gauzen. Sov. J. Quantum Electron., 16, 471-474 (1986).

[13] A. Vasil'ev, M. A. Vorontsov, A. V. Koryabin, A. F. Naumov, V. I. Shmal'gauzen. Sov. J. Quantum Electron. 19, 395-398 (1989).

[14] D. Bonaccini, G. Brusa, S. Esposito, P. Salinari, P. Stefanini. SPIE, 1334, 89-97 (1990).

[15] D. Bonaccini, G. Brusa, S. Esposito, P. Salinari, P. Stefanini, V. Biliotti. SPIE, 1543, 133-143 (1991).

[16] R. Dou, M. K. Giles. Optics Letters, 20, 1583-1585 (1995).

[17] S. R. Restaino. SPIE, 2200, 46-48 (1994).

[18] S. R. Restaino, T. Martinez, J. R. Andrews, S. W. Teare. SPIE, 4825, 41-45 (2002).

[19] P. V. Mitchell. "Innovative adaptive optics using liquid crystal light valve," Optical Society of America, (1992).

[20] G. D. Love, J. S. Fender, S. R. Restaino. Opt. And Phot. News, 6, 16-20 (1995).

[21] G. D. Love, John V. Major, Alan Purvis. Optics Letters, 19, 1170-1172 (1994).

[22] A.V. Kudryashov, J. Gonglewski, S. Browne, R. Highland. Opt. Comm. 141, 247-253 (1997).

[23] D. C. Dayton, S. L. Browne, S. P. Sandven, J. D. Gonglewski, and A. V. Kudryashov. App. Opt., 37, 5579-5589 (1998).

[24] T. L. Kelly, G. D. Love. App. Opt., 38, 1986-1989 (1999).

[25] G. T. Bold, T. H. Barnes, J. Gourlay, R. M. Sharples, T. G. Haskell. Optics Communications, 148, 323-330 (1998).

[26] J. Gourlay, G. D. Love, P. M. Birch, R. M. Sharples, and A. Purvis. Opt. Comm., 137, 17-21 (1997).

[27] D. Bonaccini, et. al. SPIE, 2000, 96-98 (1993).

[28] V. A. Dorezyuk, A. F. Naumov, V. I. Shmalgauzen. Sov. Phys. Tech. Phys., 34, 1384 (1989).

[29] W. Klaus, et. al. SPIE, 3635, 66-73 (1999).

[30] D. Dayton, J. Gonglewski, S. Restaino, J. Martin, J. Phillips, M. Hartman, S. Browne, P. Kervin, J. Snodgrass, N. Heimann, M. Shilko, R. Pohle, B. Carrion, C. Smith, D. Thiel. Optics Express, 10, 1508-1519 (2002).

[31] Quanquan Mu, Zhaoliang Cao, Lifa Hu, Yonggang Liu, Zenghui Peng, Lishuang Yao, Li Xuan. Optics Communications, 285, 896-899 (2012)

[32] Zenghui Peng, Yonggang Liu, Lishuang Yao, Zhaoliang Cao, Quanquan Mu, Lifa Hu, and Li Xuan. Optics Letters, 36, 3608–3610 (2011).

[33] Quanquan Mu, Zhaoliang Cao, Lifa Hu, Yonggang Liu, Zenghui Peng, Li Xuan. Optics Express, 18, 21687-21696 (2010)

[34] Zhaoliang Cao, Quanquan Mu, Lifa Hu, Yonggang Liu, Li Xuan. Optics Communications, 283,946-950 (2010)

[35] Zhaoliang Cao, Quanquan Mu, Lifa Hu, Xinghai Lu, and Li Xuan. Opt. Express, 17, 9330-9336 (2009).

[36] Zhaoliang Cao, Quanquan Mu, Lifa Hu, Dayu Li, Zenghui Peng, Yonggang Liu, Li Xuan. Opt. Express, 17, 2530-2537 (2009).

[37] Ran Zhang, Jun He, Zenghui Peng, Xuan Li. Chinese Physics B, 18, 2885-92 (2009).

[38] Zhaoliang Cao, Quanquan Mu, Lifa Hu, Dayu Li, Yonggang Liu, Lu Jin, Li Xuan. Opt. Express, 16, 7006-7013 (2008)

[39] Zhaoliang Cao, Quanquan Mu, Lifa Hu, Yonggang Liu, Zenghui Peng, Li Xuan. Applied Optics, 47, 1785-1789 (2008)

[40] Quanquan Mu, ZhaoLiang Cao, Dayu Li, Lifa Hu, Li Xuan. Applied Optics, 47, 4297-4301 (2008).

[41] Zhaoliang Cao, Li Xuan, Lifa Hu, Yongjun Liu, Quanquan Mu, Dayu Li. Optics Express, 13, 1059-1065 (2005).

[42] Lifa Hu, Li Xuan, Yongjun Liu, Zhaogliang Cao, Dayu Li, QuanQuan Mu. Optics Express, 12, 6403-6409 (2004).

[43] F. V. Martin, P. M. Prieto, P. Artal. J. Opt. Soc. Am. A, 15, 2552-2562 (1998).

[44] A. Awwal, B. Bauman, D. Gavel, S. S. Olivier, S. Jones, J. L. Hardy, T. Barnes, J. S. Werner. SPIE, 5169, 104-122 (2003).

[45] W. Quan, Z. Wang, G. Mu, L. Ning. Optik, 114, 1-5 (2003).

[46] X. Wang, D. Wilson, R. Muller, P. Maker, D. Psaltis. Applied Optics, 39, 6545-6555 (2000).

[47] S. Serati, J. Stockley. IEEE Aerospace Conf. Proc. 3, 1395-1402, (2002).

[48] J. Stockley, S. Serati. SPIE, 5550, 32-39 (2004).

[49] N. V. Tabiryan, S. R. Nersisyan. Applied Physics Letters, 84, 5145-5147 (2004).

[50] S. Serati, J. Stockley. SPIE, 5894, 58940K.1-58940K.13 (2005).

[51] M. Reicherter, T. Haist, E. U. Wagemann, H. J. Tiziani. Optics Letters, 24, 608-610 (1999).

[52] W. Hossack, E. Theofanidou, J. Crain, K. Heggarty, M. Birch. Optics Express, 11, 2053-2059 (2003).

[53] L. Quesada, J. Andilla, E. M. Badosa. Applied Optics, 48, 1084-1090 (2009).

[54] S. Krueger, G. Wernicke, H. Gruber, N. Demoli, M. Duerr, S. Teiwes. SPIE, 4294, 84-91 (2001).

[55] G. Wernicke, S. Kruger, H. Gruber, N. Demoli, M. Durr, S. Teiwes. SPIE, 4596, 182-190 (2001).

[56] P. Ambs, L. Bigue, E. Hueber. SPIE, 5518, 92-103 (2004).

[57] I. Moreno, A. Marquez, J. Nicolas, J. Campos, M. J. Yzuel. SPIE, 5456, 186-196 (2004).

[58] V. G. Chigrinov. SPIE, 5003, 130-137 (2003).

[59] L. Scolari, T. T. Alkeskjold, J. Riishede, A. Bjarklev, D. S. Hermann, Anawati, M. D. Nielsen, P. Bassi. Optics Express, 13, 7483-7496 (2005).

[60] J. D. Schmidt, M. E. Goda, B. D. Duncan. SPIE, 6711, 67110M.1-67110M.12 (2007).

[61] L. Hu, L. Xuan, Z. Cao, Q. Mu, D. Li, Y. Liu. Optics Express, 14, 11911-11918 (2006).

[62] K. Hirabayashi, T. Yamamoto, M. Yamaguchi. "Free space optical interconnections with liquid crystal microprism arrays," Applied Optics, Vol. 34, 2571-2580 (1995).

[63] Y. H. Lin, M. Mahajan, D. Taber, B. Wen, B. Winker. SPIE, 5892, 58920C.1-58920C.10 (2005).

[64] M. Ferstl, B. Kuhlow, E. Pawlowski. Optical Engineering, 33 1229-1235 (1994).

[65] H. Li, Z. Lu, J. Liao, Z. Weng. Acta Photonica Sinica, 29, 559-563 (2000) (In Chinese).

[66] P. Xu, X. Zhang, L. Guo, Y. Guo, et al. Acta Photonica Sinica, 16, 833-838 (1996) (In Chinese).

[67] R. Hudgin. J. Opt. Am., 67, 393-395 (1977).

[68] F. Roddier, Adaptive Optics in Astronomy (Cambridge University Press, 1999), pp. 13-15.

[69] R. K. Tyson, Principles of adaptive optics (Second Edition Academic Press 1997), pp. 71.

[70] G. D. Love, "Liquid crystal adaptive optics," in: Adaptive optics engineering handbook (R. K. Tyson, CRC, 1999).

[71] Z. Cao, L. Xuan, L. Hu, Y. Liu, Q. Mu. Opt. Express 13, 5186-5191 (2005).

[72] L. N. Thibos, A. Bradley. Optometry and Vision Science 74, 581-587 (1997).

[73] Z. Cao, Q. Mu, L. Hu, et al. Chin. Phys., 16, 1665-1671 (2007).

[74] R. J. Noll. J. Opt. Soc. Am., 66, 207-211 (1976).

[75] N. Roddier. Optical Engineering, 29, 1174-1180 (1990).

[76] Mikhail Loktev, Gleb Vdovin, Nikolai Klimov, et al. Opt. Express, 15, 2770-2778 (2007).

[77] U. Efron, J. Grinberg, P. O. Braatz, M. J. Little, P. G. Reif, R. N. Schwartz. J. Appl. Phys., 57, 1356-1368 (1985).

[78] U. Efron, S. T. Wu, J. Grinberg, L. D. Hess. Opt. Eng., 24, 111-118 (1985).

[79] N. Konforti, E. Marom, S. T. Wu. Optics Letters, 13, 251-253 (1988).

[80] V. Laude. Optics Communications, 153, 134-152 (1998).

[81] Quanquan Mu, ZhaoLiang Cao, Dayu Li, Lifa Hu, Li Xuan. Applied Optics, 47, 4297-4301 (2008).

[82] Gu, B. Winker, B. Wen, et al. Proc. SPIE, 5553, 68-82, (2004).

[83] P. Birch, J. Gourlay, G. Love, et al. Appl. Opt., 37, 2164-2169 (1998).

[84] Jakeman, E.P. Raynes Phys. Lett., 39A, 69-70 (1972).

[85] S. Gauza, H. Wang, C. Wen, S. Wu, A. Seed, R. Dabrowski, Jpn. J. Appl. Phys., 42, 3463-3466 (2004).

[86] R. Zhang, Z. Peng, Y. Liu, L. Xuan. Chinese Physics B, 18, 4380-4385 (2009).

[87] A.V. Zakharov, A. V. Komolkin, A. Maliniak. Phys. Rev. E, 59, 6802-6807 (1999).

[88] M. Fialkowski Phys. Rev. E, 58, 1955-1966 (1998).

[89] M. F. Vuks, Opt & Spectroscopy, 20, 361-368 (1966).

[90] S. Gauza, H. Wang, C. Wen, S. Wu, A. Seed, R. Dabrowski. Jpn. J. Appl. Phys., 42, 3463-3466 (2004).

[91] S. Gauza, C. Wen, B. Wu, S. Wu, A. Spadlo, R. Dabrowski. Liq. Cryst., 33, 705-710 (2006).

[92] C. O. Catanescu, S. T. Wu, L. C. Chien. Liq. Cryst., 31, 541-555 (2004).

[93] S. Gauza, A. Parish, S. Wu, A. Spadlo, R. Dabrowski. Liq. Cryst., 35, 483-488 (2008).

[94] Z. Peng, Y. Liu, L. Yao, et al. Chinese Journal of Liquid Crystal and Display, 26, 427-431 (2011) (In Chinese).

[95] Z. Peng, Y. Liu, L. Yao, Z. Cao, Q. Mu, L. Hu, X. Lu, L. Xuan, Z. Zhang. Chinese Physics Letters, 28, 094207-1-094207-3 (2011).

Chapter 6

Digital Adaptive Optics:
Introduction and Application to Anisoplanatic Imaging

Mathieu Aubailly and Mikhail A. Vorontsov

Additional information is available at the end of the chapter

1. Introduction

1.1. Conventional adaptive optics

Adaptive optics (AO) was originally developed for astronomical applications and aims to compensate the degrading effect of atmospheric turbulence on optical imaging systems performance [1]. It was later adapted to other applications such as free-space laser communication, surveillance, remote sensing, target tracking and laser weapons [2]. It also found applications in the medical field with retinal imaging [3] and potentially laser surgery. While in the later case degradations are induced by the Earth atmosphere, ocular aberrations are the limiting factor in the latter case.

Regardless of the application of interest, conventional AO systems typically perform two tasks: (1) they sense the wavefront aberrations resulting from wave propagation through the random media (e.g. the atmosphere), and (2) they compensate these aberrations using a phase conjugation approach. The components required to perform these tasks typically consist of a wavefront sensor (WFS) such as the widely-used Shack-Hartmann WFS, a wavefront corrector (WFC) – typically a deformable or segmented mirror – and a control device that computes the actuator commands sent to the WFC from the WFS data. This compensation process must be performed at speeds that match or exceed the rate of evolution of the random media – so-called real-time compensation. As a result of this requirement conventional adaptive optics systems are usually complex and often costly.

Although the conventional AO approach successfully mitigates turbulence-induced wavefront phase aberrations it presents fundamental and technological limitations.

1.2. Limitations of conventional adaptive optics

Performance of AO systems is limited by a number of factors among which wavefront correctors performance have a strong impact. First, WFC's have a limited number of degrees-of-freedom. For example, the number of control channels of a deformable mirror seldom exceeds a few tens across its aperture. This limitation affects the spatial scale of the wavefront features the WFC can compensate and prevents the system from mitigating high-order aberrations (i.e. aberrations with small spatial features). This constrain is especially critical for optical systems with aperture diameter $D \gg r_0$, where r_0 is the Fried parameter [4]. Another restrictive feature of WFC's is the limited amplitude of the wavefront phase they can compensate. This limitation prevents in parts conventional AO systems to be effective under strong (deep) turbulence conditions, which are typical for optical systems operating over long and/or near-horizontal (slant) atmospheric propagation paths. Finally, the limited temporal response of WFC's may prevent them from providing compensation at rates that exceed the rate of aberration changes.

Although technological developments have been providing WFC's with higher spatial resolution, increased dynamical range and bandwidth, an effect known as anisoplanatism which is reviewed briefly in the next section remains a fundamental limitation for adaptive optics compensation.

1.3. Anisoplanatism

Conventional AO systems typically require a reference beam (guide star) that is used to probe the atmospheric turbulence and provide an optical signal to the WFS [5]. However, the light arising from different directions within the scene does not experience the same atmospheric turbulence aberrations (propagation through volume turbulence) [6]. This causes AO performance to vary spatially across the field-of-view (FOV) with best image quality achieved for directions near the reference beam and over a small angular subtense in the order of the isoplanatic angle θ_0 [7]. The isoplanatic angle depends on the turbulence strength profile $C_n^2(z)$ where z is the altitude, and is given by

$$\theta_0 = \frac{58.1 \times 10^{-3} \lambda^{6/5}}{\left[(\sec \theta_z)^{8/3} \int_0^L C_n^2(z) z^{5/3} dz \right]^{3/5}}, \tag{1}$$

where θ_z is the Zenith angle of observation and λ is the wavelength [4]. Even under conditions of weak turbulence θ_0 is usually small and remains in the order of a few microradians to a few tens of microradians. The isoplanatic angle is especially narrow for near-ground and near-horizontal propagation paths (i.e. high and nearly constant C_n^2 values). Anisoplanatism degrades the performance of AO systems as the angular separation θ (known as field angle) between the reference beam and points on the object increases [8].

A number of techniques have been developed to mitigate the effect of anisoplanatism such as using multiple WFS's and WFC's located in optical conjugates of planes at various distances along the light of sight – an approach referred to as multi-conjugate AO (MCAO) [9-11]. Using multiple guide stars distributed within the field-of-view has also been explored [12]. Although these approaches have been shown to be effective, they both result in significant increase of system complexity and cost. Post-processing techniques have been investigated but they usually assume knowledge of the point spread function (PSF) for several values of the field angle θ [13].

In the remainder of this chapter we introduce an alternative approach to conventional adaptive optics – referred to as digital adaptive optics (DAO) – which alleviates the need for physical WFC devices and their corresponding real-time control hardware, and relieves the system from the limitations associated to them (see section 1.2). In section 2 we present the approach used in DAO systems and discuss their limitations. The DAO technique is then applied to an anisoplanatic imaging scenario and results of numerical analysis are presented in section 3. Finally section 4 draws conclusions.

2. Digital adaptive optics

2.1. General approach

The notional schematic in Figure 1 shows the sequential steps required for obtaining a compensated image using the digital adaptive optics approach. Two major steps of the process are as follow:

Step 1: Optical field measurement

The front-end of the DAO system consists of an optical reducer and an optical sensor referred to as complex-field sensor (CFS). The CFS provides simultaneous measurements of the optical field wavefront phase and intensity distributions in its pupil plane, denoted $\varphi(\mathbf{r})$ and $I(\mathbf{r})$ respectively. This sensor is referred to as a complex field sensor since the complex amplitude of optical field $A(\mathbf{r})$ can be represented in the form $A(\mathbf{r}) = |A(\mathbf{r})| \exp\{j\varphi(\mathbf{r})\}$, where $|A(\mathbf{r})| = I^{1/2}(\mathbf{r})$, and both phase $\varphi(\mathbf{r})$ and amplitude $|A(\mathbf{r})|$ functions can be obtained from the sensor measurements. The reducer is used for re-imaging of the DAO system pupil onto the CFS pupil so that $A(\mathbf{r}) \approx A_{in}(M\mathbf{r})$ where $A_{in}(\mathbf{r})$ denotes the complex amplitude of the optical field entering the DAO system. The term M is a scaling factor associated with the beam reducer and $\mathbf{r} = \{x, y\}$ designates a coordinate vector in the system pupil plane. To simplify notation, we assume $M = 1$ and $A(\mathbf{r}) \approx A_{in}(\mathbf{r})$. Section 2.3 provides details about complex field sensing techniques for DAO systems.

As a result of propagation through atmospheric turbulence the wavefront phase $\varphi_{in}(\mathbf{r})$ received at the DAO system's pupil (and measured by the CFS sensor) can be separated into two components:

$$\varphi_{in}(\mathbf{r}) = \varphi_{obj}(\mathbf{r}) + \varphi_{turb}(\mathbf{r}) \qquad (2)$$

where $\varphi_{obj}(\mathbf{r})$ is the phase component related to the object of interest (scene) and $\varphi_{turb}(\mathbf{r})$ is the turbulence-induced phase term which needs to be compensated. The second step of the DAO process aims to (1) compensate phase aberrations $\varphi_{turb}(\mathbf{r})$ which degrade the quality of the images produced by the system and (2) preserve phase $\varphi_{obj}(\mathbf{r})$ which is used to synthesize a compensated image.

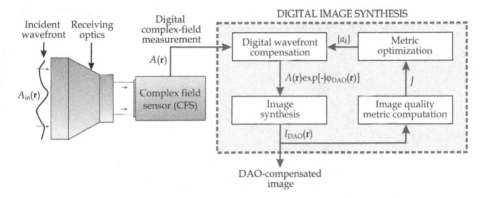

Figure 1. Notional schematic of a digital adaptive optics system.

Step 2: Digital image synthesis

In this second step of the DAO technique, a digital processing technique is used to synthesize a compensated image from the complex-field measurement $A(\mathbf{r})$ provided by the CFS (see step 1). As illustrated in Fig. 1 this step includes compensation of the measured complex field $A(\mathbf{r})$ using phase function $\varphi_{DAO}(\mathbf{r})$ and results in compensated field $A_{DAO}(\mathbf{r}) = A(\mathbf{r})\exp\{-j\varphi_{DAO}(\mathbf{r})\}$. This phase conjugation step using numerical phase function $\varphi_{DAO}(\mathbf{r})$ can be regarded as the digital equivalent of phase conjugation using a wavefront corrector such as a deformable mirror or a liquid crystal phase modulator which are used in conventional AO systems. We refer to this step as digital wavefront compensation (DWFC).

The compensated complex-field $A_{DAO}(\mathbf{r})$ is then used to synthesize image $I_{DAO}(\mathbf{r})$. Image quality of $I_{DAO}(\mathbf{r})$ hence depends on phase $\varphi_{DAO}(\mathbf{r})$ applied at the DWFC step. The quality of image $I_{DAO}(\mathbf{r})$ is assessed by computing an image quality (sharpness) metric J, and an algorithm is used to optimize metric J, leading to an image $I_{DAO}(\mathbf{r})$ with improved quality. Details about the image formation and optimization process are presented in section 2.4.

2.2. Comparison between conventional and digital AO system operations

For a conventional AO system to operate successfully wavefront compensation (conjugation) is required to be performed during time $\tau_{AO} < \tau_{at}$ where τ_{at} is the characteristic time of atmospheric turbulence change (i.e. under "frozen" turbulence conditions). Such a system is referred to as *real-time* system and its temporal response results of the combination of the individual response time of each element of the AO feedback loop shown in Fig. 2(a) so that:

$$\tau_{AO} = \tau_{WFC} + \tau_{WFS} + \tau_{cont} \tag{3}$$

where τ_{WFC}, τ_{WFS} and τ_{cont} correspond respectively to the temporal response of the WFC, WFS and controller devices. Bandwidth requirements hence apply to each element of the feedback loop and drive in part the cost of AO systems.

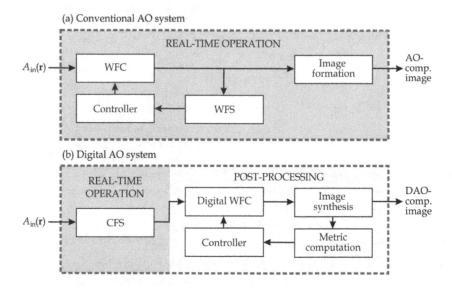

Figure 2. Block diagram identifying keys components of (a) a conventional AO system and (b) a digital AO system. While conventional AO requires both wavefront sensing and wavefront compensation to be performed in real-time, digital AO requires only complex-field sensing to be realized in real-time. Subsequent digital image formation and compensation can be performed as a post-processing step.

Although digital AO systems are based on the same principle of phase conjugation than conventional AO, compensation is implemented in a difference manner. DAO systems employ numerical techniques and do not use physical wavefront corrector devices such as deformable or segmented mirrors. This has the advantage of alleviating the need for WFC's and their real-time control hardware, two elements that impacts significantly the cost and complexity of conventional AO systems.

In a DAO system measurements of the input complex-field $A_{in}(\mathbf{r})$ are performed during $\tau_{CFS} < \tau_{at}$ (i.e. in real-time) while image formation and compensation of turbulence-induced phase aberrations are performed as *post-processing* steps as illustrated in Fig. 2(b). Since the post-processing step is not required to be performed at high speed, it can be achieved using standard computation techniques such as a PC, which simplifies implementation of DAO systems. The time delay associated with image synthesis and compensation using such techniques may be suitable for some applications. However, for applications where real-time DAO operation is critical, the post processing step may be implemented on a dedicated high-speed hardware.

2.3. Wavefront sensing techniques for DAO systems

In conventional AO systems the spatial resolution of the wavefront sensor output (i.e. the spacing between data points) is related to the spatial resolution of the wavefront corrector (e.g. spacing between deformable mirror actuators). Sensing of the incoming wavefront aberrations with high spatial frequency does not provide better AO performance if the corrector device is unable to match this spatial resolution. In the other hand image quality in DAO systems is directly related to the spatial resolution of the wavefront measurement. DAO systems hence require high resolution wavefront sensing capabilities.

Although the resolution yielded by wavefront sensors typically used in adaptive optics such as Shack-Hartmann [1,14] or curvature sensors [15,16] does not exceed a few tens to a couple hundred data points across the system's aperture, a number of wavefront sensing techniques capable of providing high resolution outputs exist. Among them some are potentially suitable for DAO systems including: phase retrieval from sets of pupil and focal plane intensity distributions [17-19], phase diversity [20,21], schlieren techniques and phase contrast techniques [22,23] such as the Zernike filter [24-26] and the Smartt point-diffraction interferometer [27-30]. Approaches based on holographic recording of the wavefront have also been used successfully [31-33]. The recently developed sensor referred to as multi-aperture phase reconstruction (MAPR) sensor [34] uses a hybrid approach between the Shack-Hartmann and Gerchberg-Saxton [17] techniques to provide high-resolution measurements and is also a candidate for DAO system implementation.

A growing number of applications now require operation over near-horizontal or slant atmospheric paths. These propagations scenarios are characterized by moderate to strong intensity scintillation [35-37]. This means that in addition to high resolution requirements, robustness to high scintillation levels is a critical criterion for selecting sensors suitable for DAO applications. Another important criterion is the computational cost of the wavefront reconstruction algorithm as it impacts the speed of operation of the sensor. In this regard the MAPR sensor might be suitable for DAO applications since it is capable of providing high-resolution measurements under conditions of strong intensity scintillation (so called scintillation-resistant) and in the presence of branch points [38,39]. It yields an average Strehl ratio exceeding 0.9 for scintillation index values $\sigma_I^2 \leq 1.25$ and $D/r_0 \leq 8$, and 0.8 for $\sigma_I^2 \leq 1.75$ and $D/r_0 \leq 12$ and reconstruction is computationally efficient as a result of the parallel nature of the algorithm.

However, as the selection of wavefront sensing techniques suitable for DAO-based imaging requires further investigations, we assume in the remainder of this chapter that the complex amplitude of the incident optical field $A_{in}(\mathbf{r})$ (see Fig. 1) is known and that $A(\mathbf{r}) = A_{in}(\mathbf{r})$. DAO image synthesis and compensation techniques based on measurement $A(\mathbf{r})$ are provided in the next section.

2.4. Anisoplanatic image synthesis and compensation

As a result of anisoplanatism (see section 1.3) image quality varies significantly across the field-of-view of the system and image compensation based on phase conjugation (section 2.1) is effective only over a small angular extent with size related to the isoplanatic angle θ_0. An approach for performing efficient DAO compensation is to apply the technique locally over image regions that are nearly isoplanatic. We present in this section a block-by-block (mosaic) post-processing technique in which the DAO approach is applied sequentially to regions (blocks) Ω_j and the resulting image consists in the combination of compensated image regions Ω_j into a single image corresponding to the entire FOV (region Ω). Consider an image region Ω_j defined by function

$$M_{\Omega_j}(\mathbf{r}) = \exp\left[\left(-\frac{|\mathbf{r} - \mathbf{r}_j|^2}{2\omega_\Omega^2}\right)^8\right], \tag{4}$$

where $\mathbf{r}_j$ defines the position of the jth image region and ω_Ω denoted its size. Image synthesis and DAO compensation over region Ω_j is performed based on the measurement $A(\mathbf{r})$ of the optical field in the pupil of the system and consists of the following steps:

Step 1: Digital wavefront correction

A wavefront corrector phase function $\varphi_{DAO}(\mathbf{r})$ is represented as

$$\varphi_{DAO}(\mathbf{r}) = \sum_{k=1}^{N_{DAO}} a_k S_k(\mathbf{r}), \tag{5}$$

where $\{S_k(\mathbf{r})\}$ is a set of response functions for the digital wavefront corrector, $\mathbf{a} = \{a_k\}$ is the vector of commands sent to the DWFC, and N_{DAO} is the number of control channels of the DWFC. The corrector phase function $\varphi_{DAO}(\mathbf{r})$ is then applied to the complex field amplitude $A(\mathbf{r})$ provided by the CFS and results in compensated field $A_{DAO}(\mathbf{r})$ given by

$$A_{DAO}(\mathbf{r}) = A(\mathbf{r})\exp\left[-j\varphi_{DAO}(\mathbf{r})\right]. \tag{6}$$

Step 2: Image synthesis

The compensated field $A_{DAO}(\mathbf{r})$ in Eq. (6) is used for synthesis of the DAO compensated image $I_{DAO}(\mathbf{r})$. The DAO image is computed using the Fresnel approximation as follows [25]:

$$I_{DAO}(\mathbf{r}) = \left| \frac{1}{\lambda Z} \int_{-\infty}^{+\infty} A_{DAO}(\mathbf{r}')\exp\left[-j\frac{k}{2}\left(\frac{|\mathbf{r}'|^2}{F} - \frac{|\mathbf{r}-\mathbf{r}'|^2}{I_i}\right)\right]d\mathbf{r}' \right|^2, \tag{7}$$

where k is the wave number, F is the focal length of the digital lens and $L_i = (1/F - 1/L)^{-1}$ is the distance between the digital lens plane and the image plane. The term L denotes the distance between the lens and the object plane of interest.

Step 3: Local image quality metric computation

The quality of the synthesized image $I_{DAO}(\mathbf{r})$ in Eq. (7) depends on the command vector $\mathbf{a}$ applied to the digital WFC [Eq. (5)]. Vector $\mathbf{a}$ can be considered as a parameter controlling the quality of image $I_{DAO}(\mathbf{r})$. Improving image quality in the region Ω_j can be achieved by optimizing a sharpness metric J_{Ω_j} given by

$$J_{\Omega_j}(\mathbf{a}) = \frac{\int I_{DAO}^2(\mathbf{r})M_{\Omega_j}(\mathbf{r})d\mathbf{r}}{\int I_{DAO}(\mathbf{r})M_{\Omega_j}(\mathbf{r})d\mathbf{r}}, \tag{8}$$

Where $M_{\Omega_j}(\mathbf{r})$ is the function defining region Ω_j [see Eq. (4)]. Note that although the intensity-squared sharpness metric is commonly used, other criteria could be used to assess image quality such as a gradient-based metric or the Tenengrad criterion.

Step 4: Image quality metric optimization

Optimization of metrics J_{Ω_j} can be achieved using various numerical techniques. We consider for example metric optimization based on the stochastic parallel gradient descent (SPGD) control algorithm [40]. In accordance with this algorithm command vector $\mathbf{a}$ update rule is given at each iteration n by the following procedure:

$$a_k^{(n+1)} = a_k^{(n)} + \gamma^{(n)}\delta J_{\Omega_j}^{(n)}\delta a_k^{(n)} \text{ for } k = 1, \ldots, N_{DAO}, \tag{9}$$

where $\gamma^{(n)}>0$ is a gain coefficient, $\delta J_{\Omega_j}^{(n)}$ is the metric response to small-amplitude random perturbations of control vector $\{\delta a_k^{(n)}\}$ applied simultaneously to all N_{DAO} DWFC control channels. The control channel updates are repeated until convergence of vector $\mathbf{a}$ toward a small vicinity of the stationary state. The number of iterations N_{it} required for convergence is defined from the common criterion

$$\varepsilon\left(n=N_{it}\right)=\frac{\left|J_{\Omega_j}^{(n)} - J_{\Omega_j}^{(n-1)}\right|}{J_{\Omega_j}^{(n)}}\leq\varepsilon_0\ll 1, \tag{10}$$

and the resulting control vector for compensation of region Ω_j, denoted $\mathbf{a}_{\Omega_j}$, corresponds to the vector obtained at the last iteration: $\mathbf{a}_{\Omega_j}=\mathbf{a}^{(n=N_{it})}$. Optimization over the entire image region Ω can be achieved by repeating sequentially steps 1 through 4 for each region Ω_j and results in a set of control vectors $\{\mathbf{a}_{\Omega_j}\}$.

In an ideal compensation scenario the DWFC phase $\varphi_{DAO}(\mathbf{r})$ resulting from the optimization process would compensate exactly the turbulence-induced phase aberration so that $\varphi_{DAO}(\mathbf{r})=\varphi_{turb}(\mathbf{r})$ [see Eq. (2)]. In this ideal case only phase information related to the object being imaged remains in the compensated field: $\arg\left[A_{DAO}(\mathbf{r})\right]=\varphi_{obj}(\mathbf{r})$ and leads to optimal image quality.

3. Performance analysis

In this section performance of DAO systems is analyzed using a numerical simulation and results for various system configurations and turbulence strengths are discussed.

3.1. Numerical model

Performance was evaluated from an ensemble of digitally-generated random complex-fields used as input optical waves to the DAO system. For each realization of the input field the complex amplitude in the DAO system pupil plane $A_{in}(r)$ (see Fig. 1) was obtained using the conventional split-operator approach for simulating wave optics propagation through a volume of atmospheric turbulence [41]. At the beginning of the propagation path ($z=0$) we used a monochromatic optical field with complex amplitude $A_{prop}(\mathbf{r}, z=0)=I_{obj}^{1/2}(\mathbf{r})\exp\left[j\varphi_{surf}(\mathbf{r})\right]$, where $I_{obj}(\mathbf{r})$ is the intensity distribution of the object being imaged and $\varphi_{surf}(\mathbf{r})$ is a random phase function uniformly distributed in $[-\pi, \pi]$ and δ-correlated in space. The term $\varphi_{surf}(\mathbf{r})$ is used to model the "optically rough" surface of the object. The optical field complex amplitude at the end of the propagation path ($z=L$) was utilized as the DAO system input field: $A_{in}(\mathbf{r})=A_{prop}(\mathbf{r}, z=L)$. Optical inhomogeneities along the propagation path were modeled with a set of 10 thin random phase screens corresponding to the Kolmogorov turbulence power spectrum. We considered a horizontal propagation scenario so the phase screens were equally spaced along the propagation path and their impact (i.e. turbulence strength) was characterized by a constant ratio D/r_0 where D is the diameter of the DAO system aperture and r_0 is the characteristic Fried parameter for plane waves. By modifying ratio D/r_0 one can control the strength of input field phase aberrations. In the numerical simulations, D/r_0 ranged from zero (i.e. free-space propagation) to 10. Figures 3(a)-3(d) show examples of the input field intensity and phase distributions that are obtained using the technique described above for $D/r_0=4$ [(a) and (b)] and for $D/r_0=8$ [(c) and (d)]. Note that the phase distributions in Figs. 3(b) and 3(d) contain phase discontinuities (branch points). Images were obtained for a propagation distance $L=0.05L_{diff}$ where $L_{diff}=k(D/2)^2$ is the diffractive distance, $k=2\pi/\lambda$ is the wave number, and λ is the imaging wavelength.

The strength of the input field intensity scintillations was characterized by the aperture-averaged scintillation index σ_I^2 given by

$$\sigma_I^2 = \frac{1}{S} \int \left\{ \frac{\langle [I(\mathbf{r})]^2 \rangle}{\langle I(\mathbf{r}) \rangle^2} - 1 \right\} d^2\mathbf{r}, \tag{11}$$

where $I(\mathbf{r}) = |A_{in}(\mathbf{r})|^2$ and S is the aperture area of the DAO system. Here $<>$ denotes averaging over an ensemble of input fields corresponding to statistically independent realizations of the phase screens as well as the object roughness phase function $\varphi_{surf}(\mathbf{r})$. In the numerical simulations, ensemble averaging was performed over 100 sets of phase screens and object phases. Both intensity distributions in Figs. 3(a) and 3(c) are characterized by strong scintillations with an average scintillation index value of 1.

In the DAO system, digital wavefront compensation function $\varphi_{DAO}(\mathbf{r})$ is computed as a weighted sum of response functions $S_k(\mathbf{r})$ [see Eq. (5)]. In our numerical simulation these response functions are taken as scaled Zernike polynomials: $S_k(\mathbf{r})=Z_k(Q\mathbf{r})$, where $Z_k(\mathbf{r})$ is the Zernike polynomial with index k and Q is a scaling factor. Since Zernike polynomials are defined onto the unit circle the term Q is chosen as $Q=D/2$ so that response functions $S_k(\mathbf{r})$ are defined over the aperture area of the DAO system. In the numerical analysis we used $N_{DAO}=36$ Zernike polynomials corresponding to the first 8 Zernike modes. It should be noted that the first Zernike polynomial (piston) does not impact image quality and does not need to be included in the construction of $\varphi_{DAO}(\mathbf{r})$. Similarly, Zernike polynomials with index $k=2$ and $k=3$ (tip and tilt) do not have an influence on image quality or its metric. However, tip-tilt wavefront aberrations in the pupil plane cause a global shift of the intensity distribution in the image plane. In order to compensate for this, image alignment is performed in our numerical simulations by mean of a conventional registration technique [42] using an ensemble average image as a reference.

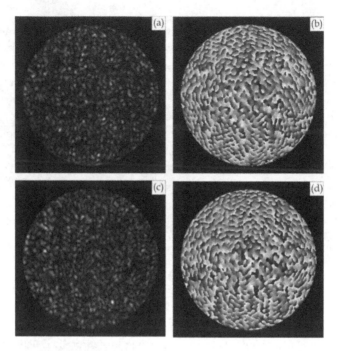

Figure 3. Intensity (left column) and phase (right column) distributions of the computer-generated complex field $A_{in}(\mathbf{r})$ used as input of the DAO system for $D / r_0 = 4$ [(a),(b)] and $D / r_0 = 8$ [(c),(d)]. Phase distributions are shown in a 2π range [between $-\pi$ (black) and $+\pi$ (white)].

The quality of the DAO-compensated image $I_{DAO}(\mathbf{r})$ is assessed using a sharpness metric defined as

$$J = \frac{\int I_{DAO}^2(\mathbf{r})d\mathbf{r}}{\int I_{dl}^2(\mathbf{r})d\mathbf{r}} , \qquad (12)$$

Where $I_{dl}(\mathbf{r})$ is the image that would be obtained in the absence of atmospheric turbulence (i.e. diffraction-limited image). In the case of an ideal compensation the turbulence-induced wavefront aberrations are fully corrected by the DWFC [$\varphi_{DAO}(\mathbf{r}) = \varphi_{turb}(\mathbf{r})$] and $J = 1$.

3.2. DAO system performance

In this section we analyze results of image synthesis and compensation using the DAO approach described in section 2 using the numerical model presented in section 3.1. In order to illustrate the effect of anisoplanatism we first consider DAO performance over an image region Ω_j that is nearly isoplanatic in size. Later we consider processing of the entire image region Ω (anisoplanatic conditions). Figure 4 shows intensity distribution $I_{DAO}(\mathbf{r})$ and cross-sections for an object with intensity distribution consisting of point sources arranged in a 7-by-7 array. The resulting intensity distributions hence correspond to the average DAO-compensated PSF of the system for different values of the field angle θ, and is showed for $D/r_0=4$ [Figs. 4(a) and 4(b)] and for $D/r_0=8$ [Figs. 4(c) and 4(d)]. As a result of anisoplanatism the PSF is space-varying and its distribution broadens significantly as the angular separation θ with the metric region Ω_j increases. Also, the degradation occurs more rapidly with respect to angle θ as turbulence strength increases [compare cross-sections in Figs. 4(b) and 4(d)]. Results were obtained by averaging the resulting images $I_{DAO}(\mathbf{r})$ for 100 realizations of the random phase screens.

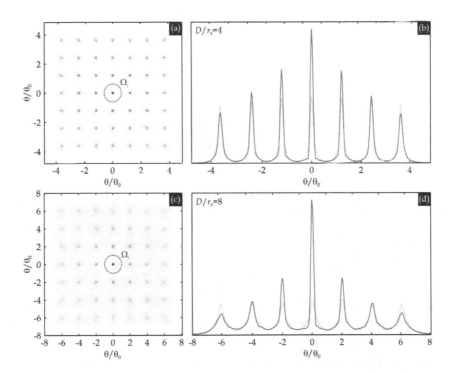

Figure 4. Average DAO-compensated intensity distributions (left column) for an array of 7-by-7 point sources for $D/r_0=4$ (top row) and $D/r_0=8$ (bottom row). The column on the right depicts the average uncompensated (dotted curve) and DAO-compensated (solid curve) cross-sections of the PSF's through the center of region Ω_j. Image compensation is based on optimizing image quality metric J_j defined over region Ω_j. Region Ω_j has a diameter w_Ω of approximately $1.2\theta_0$ and $2\theta_0$ for $D/r_0=4$ and $D/r_0=8$ respectively (see circled areas).

As described in section 2.4 compensation is based on the optimization of metric J_j using an iterative SPGD algorithm. Figure 5 shows average convergence curves for metric J_j as a function of the number of SPGD iterations performed. The two curves displayed correspond to the optimization of image regions Ω_j shown in Figure 4 for $D/r_0=4$ and $D/r_0=8$. In both cases the metric value approximately doubled during the optimization process.

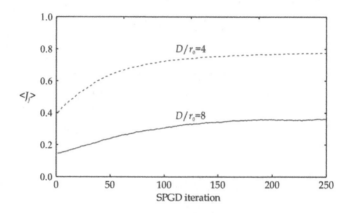

Figure 5. Average convergence curves for metric J_j. The curves displayed correspond to optimization of image region Ω_j shown in Fig. 4 for $D/r_0=4$ and $D/r_0=8$.

We consider now the performance of DAO systems under anisoplanatic conditions. Figures 6 and 7 display images prior and after DAO compensation for the point source object used in Figure 4 and for an USAF resolution chart respectively. Results are shown for two turbulence conditions: $D/r_0=4$ and $D/r_0=8$ and correspond to a full field-of-view of approximately $10\theta_0$ and $16\theta_0$. In both Figures significant image quality improvements can be observed with an increase of the metric J of 70% (point source object) and 49% (USAF object) for $D/r_0=4$, and 123% and 33% for $D/r_0=8$.

The performance of the DAO compensation technique with respect to turbulence strength is illustrated in Fig. 8 for a ratio D/r_0 in the range $[0;10]$. Within that range the average image quality metric $\langle J \rangle$ after compensation always exceeds significantly the metric value prior processing. Image quality improvements become especially important for large D/r_0. For example, for $D/r_0=10$ the metric grew by a factor of nearly 3. Plots showed in Fig. 8 and in the remainder of the chapter as shown for the object consisting of an array of point sources as depicted in Fig. 6.

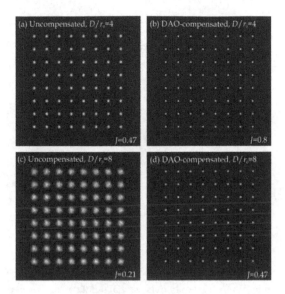

Figure 6. Image of an array of point sources prior and after anisoplanatic DAO compensation for $D/r_0=4$ (top row) and $D/r_0=8$ (bottom row).

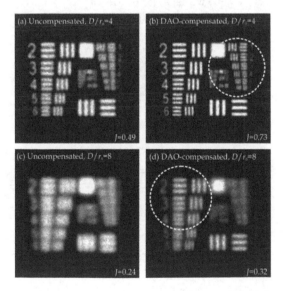

Figure 7. Image of an USAF resolution chart prior and after anisoplanatic DAO compensation for $D/r_0=4$ and $D/r_0=8$. Note that the DAO compensation process reveals image details as pointed out by the circled areas.

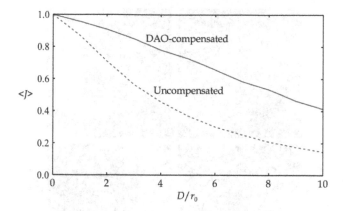

Figure 8. Average image quality metric $\langle J \rangle$ as function of the turbulence strength characterized by the ratio D / r_0 with and without DAO compensation.

As mentioned in section 1.3 anisoplanatism causes compensation approaches based on single phase conjugation to be effective only over a small angle of size related to the isoplanatic angle θ_0. As a result the block-by-block processing technique used here performs differently as the size ω_Ω of blocks Ω_j changes. As shown in Fig. 9 DAO performance degrades noticeably as ω_Ω increases. For example we have $\langle J \rangle = 0.73$ for $\omega_\Omega = 1.8\theta_0$ and $\langle J \rangle = 0.44$ for $\omega_\Omega = 14\theta_0$. This illustrates the efficiently of the block-by-block processing technique to mitigate the anisoplanatic effect.

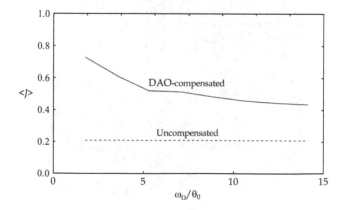

Figure 9. Average image quality metric $\langle J \rangle$ as a function of the diameter ω_Ω of region Ω_j used for metric computation.

Finally the influence of the number of Zernike polynomials N_{DAO} compensated [see Eq. (5)] is shown in Fig. 10 for two turbulence strength levels: $D/r_0=4$ and $D/r_0=8$. Image quality improvements become negligible beyond a threshold for parameter N_{DAO}. This threshold is in the range of $[10;20]$ for $D/r_0=4$ and $[15;25]$ for $D/r_0=8$.

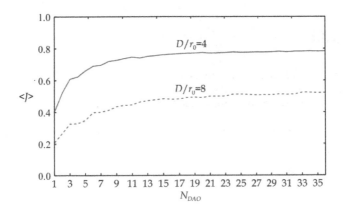

Figure 10. Average image quality metric $\langle J \rangle$ as a function of the degree of DAO compensation N_{DAO}.

4. Conclusion

We introduced in this chapter an alternative technique to conventional adaptive optics imaging schemes which we refer to as digital adaptive optics. The technique consists in a two-step process. First, an optical sensor provides a measurement of the wave's complex-amplitude (intensity and phase distributions) in the pupil of the imaging system. This differs from the conventional AO approach in which typically only the wavefront is sensed. Second, digital post-processing algorithms are applied to the complex-field measurements in order to synthesize an image and mitigate the effect of atmospheric turbulence. This final step is based on the optimization of an image quality metric and compensation of the wavefront aberrations is performed in a numerical manner. While the conventional AO approach compensates aberrations in real-time the DAO operates as a post-processing scheme. DAO systems has the advantage of requiring simpler and less costly implementations since they do not require opto-mechanical wavefront correctors and their real-time control hardware, but this also means they are primarily suited for applications that do not require real-time operation.

Performance of DAO systems was evaluated by mean of a numerical analysis. The analysis revealed the DAO approach can significantly improve image quality even in strong turbulence conditions. The block-by-block processing technique presented was shown to be effective for image synthesis and compensation under anisoplanatic scenarios. The influence of the block size on DAO performance was showed to enhance performance as the block size decreases and nears values of the isoplanatic angle. Finally, increasing the degree of DAO compensation (i.e. number of Zernike coefficients compensated) was showed to benefit performance up to a threshold value which depends on the turbulence strength.

Author details

Mathieu Aubailly[1*] and Mikhail A. Vorontsov[2]

*Address all correspondence to: mathieu@umd.edu

1 Intelligent Optics Laboratory, Institute for Systems Research, University of Maryland, College Park, Maryland, USA

2 Intelligent Optics Laboratory, School of Engineering, University of Dayton, Dayton, Ohio, USA

References

[1] J. W. Hardy, *Adaptive Optics for Astronomical Telescopes* (Oxford University, 1998).

[2] R. Tyson, *Principles of Adaptive Optics*, Third Edition (CRC Press, 2010).

[3] J. Porter, H. Queener, J. Lin, K. Thorn, A. A. S. Awwal, *Adaptive Optics for Vision Science: Principles, Practices, Design and Applications* (Wiley, 2006).

[4] M. C. Roggemann and B. M. Welsh, *Imaging Through Turbulence* (CRC-Press, 1996).

[5] B. M. Welsh and C. S. Gardner, "Performance analysis of adaptive-optics systems using laser guide stars and slope sensors," J. Opt. Soc. Am. A 6, 1913-1923 (1989).

[6] B. M. Welsh and L. A. Thompson, "Effects of turbulence-induced anisoplanatism on the imaging performance of adaptive-astronomical telescopes using laser guide stars," J. Opt. Soc. Am. A 8, 69–80 (1991).

[7] D. L. Fried, "Anisoplanatism in adaptive optics," J. Opt. Soc. Am. 72, 52–61 (1982).

[8] J. Christou, E. Steinbring, S. Faber, D. Gavel, J. Patience, and E. Gates, "Anisoplanatism within the isoplanatic patch," Am. Astron. Soc. 34, 1257 (2002).

[9] D. C. Johnston and B. M. Welsh, "Analysis of multiconjugate adaptive optics," J. Opt. Soc. Am. A 11, 394–408 (1994).

[10] B. L. Ellerbroek, "First-order performance evaluation of adaptive optics systems for atmospheric turbulence compensation in extended field of view astronomical telescopes," J. Opt. Soc. Am. A 11, 783–805 (1994).

[11] T. Fusco, J.-M. Conan, G. Rousset, L. M. Mugnier, and V. Michau, "Optimal wavefront reconstruction strategies for multiconjugate adaptive optics," J. Opt. Soc. Am. A 18, 2527–2538 (2001).

[12] M. Lloyd-Hart, C. Baranec, N. M. Milton, M. Snyder, T. Stalcup, and J. R. P. Angel, "Experimental results of ground-layer and tomographic wavefront reconstruction from multiple laser guide stars," Opt. Express 14, 7541-7551 (2006).

[13] M. Aubailly, M. C. Roggemann, and T. J. Schulz, "Approach for reconstructing anisoplanatic adaptive optics images," Appl. Opt. 46, 6055-6063 (2007).

[14] J. D. Barchers, D. L. Fried, and D. J. Link, "Evaluation of the performance of Hartmann sensors in strong scintillation," Appl. Opt. 41, 1012–1021 (2002).

[15] F. Roddier, "Curvature sensing and compensation: a new concept in adaptive optics," Appl. Opt. 27, 1223–1225 (1988).

[16] G. Rousset, "Wave-front sensors," in Adaptive Optics in Astronomy, F. Roddier, ed. (Cambridge University, 1999), pp. 91–130.

[17] R. W. Gerchberg and W. O. Saxton, "A practical algorithm for the determination of phase from image and diffraction plane pictures," Optik 35, 237–246 (1972).

[18] J. R. Fienup, "Phase retrieval algorithms: a comparison," Appl. Opt. 21, 2758-2769 (1982).

[19] V. Yu. Ivanov, V. P. Sivokon, and M. A. Vorontsov, "Phase retrieval from a set of intensity measurements: theory and experiment," J. Opt. Soc. Am. A 9, 1515-1524 (1992).

[20] R. A. Gonsalves, "Phase retrieval from modulus data," J. Opt. Soc. Am. 66, 961–964 (1976).

[21] R. G. Paxman and J. R. Fienup, "Optical misalignment sensing and image reconstruction using phase diversity," J. Opt. Soc. Am. A 5, 914–923 (1988).

[22] M. A. Vorontsov, E. W. Justh, and L. A. Beresnev, "Adaptive optics with advanced phase-contrast techniques. I. High-resolution wave-front sensing," J. Opt. Soc. Am. A 18, 1289-1299 (2001).

[23] E. W. Justh, M. A. Vorontsov, G. W. Carhart, L. A. Beresnev, and P. S. Krishnaprasad, "Adaptive optics with advanced phase-contrast techniques. II. High-resolution wave-front control," J. Opt. Soc. Am. A 18, 1300-1311 (2001).

[24] F. Zernike, "How I discovered phase contrast," Science 121, 345–349 (1955).

[25] J. W. Goodman, Introduction to Fourier Optics (McGraw-Hill, New York, 1996).

[26] S. A. Akhmanov and S. Yu. Nikitin, Physical Optics (Clarendon, Oxford, UK, 1997).

[27] R. N. Smartt and W. H. Steel, "Theory and application of point-diffraction interferometers," Jpn. J. Appl. Phys. 14, 351–356 (1975).

[28] P. Hariharan, ed., *Selected Papers on Interferometry* (SPIE Optical Engineering Press, Bellingham, Wash., 1991).

[29] R. Angel, "Ground-based imaging of extrasolar planets using adaptive optics," Nature 368, 203–207 (1994).

[30] K. Underwood, J. C. Wyant, and C. L. Koliopoulos, "Selfreferencing wavefront sensor," in Wavefront Sensing, N. Bareket and C. L. Koliopoulos, eds., Proc. SPIE 351, 108–114 (1982).

[31] J. C. Marron, R. L. Kendrick, N. Seldomridge, T. D. Grow, and T. A. Höft, "Atmospheric turbulence correction using digital holographic detection: experimental results," Opt. Express 17, 11638-11651 (2009).

[32] A. E. Tippie and J. R. Fienup, "Multiple-plane anisoplanatic phase correction in a laboratory digital holography experiment," Opt. Lett. 35, 3291-3293 (2010).

[33] N. J. Miller, J. W. Haus, P. F. McManamon, D. Shemano, "Multi-aperture coherent imaging," Proc. SPIE 8052 (2011).

[34] M. Aubailly and M. A. Vorontsov, "Scintillation resistant wavefront sensing based on multi-aperture phase reconstruction technique," J. Opt. Soc. Am. A 29, 1707-1716 (2012).

[35] V. U. Zavorotnyi, "Strong fluctuations of the wave intensity behind a randomly inhomogeneous layer," Radiophys. Quantum Electron. 22, 352–354 (1979).

[36] M. C. Rytov, Yu A. Kravtsov, and V. I. Tatarskii, eds., Principles of Statistical Radiophysics 4, *Wave Propagation Through Random Media* (Springer-Verlag, 1989).

[37] L. C. Andrews, R. L. Phillips, C. Y. Hopen, and M. A. Al-Habash, "Theory of optical scintillation," J. Opt. Soc. Am. A 16, 1417–1429 (1999).

[38] D. L. Fried, "Branch point problem in adaptive optics," J. Opt. Soc. Am. A 15, 2759–2768 (1998).

[39] D. L. Fried and J. L. Vaughn, "Branch cuts in the phase function," Appl. Opt. 31, 2865–2882 (1992).

[40] M. A. Vorontsov and V. P. Sivokon, "Stochastic parallel-gradient-descent technique for high-resolution wave-front phase-distortion correction," J. Opt. Soc. Am. A 15, 2745–2758 (1998).

[41] J. A. Fleck, J. R. Morris, and M. D. Feit, "Time-dependent propagation of high energy laser beams through the atmosphere," Appl. Phys. 10, 129–160 (1976).

[42] E. De Castro and C. Morandi, "Registration of translated and rotated images using finite Fourier transforms," IEEE Trans. Pattern Analysis and Machine Intelligence 9, 700–703 (1987).

Optical and Atmospheric Effects

Chapter 7

Adaptive Optics and Optical Vortices

S. G. Garanin, F. A. Starikov and Yu. I. Malakhov

Additional information is available at the end of the chapter

1. Introduction

The achievement of minimal angular divergence of a laser beam is one of the most important problems in laser physics since many laser applications demand extreme concentration of radiation. Under the beam formation in the laser oscillator or amplifier with optically inhomogeneous gain medium and optical elements, the divergence usually exceeds the diffraction limit, and the phase surface of the laser beam differs from the plane surface. However, even if one succeeds in realizing the close-to-plane radiation wavefront at the laser output, the laser radiation experiences increasing phase disturbances under the propagation of the beam in an environment with optical inhomogeneities (atmosphere). These disturbances appear with the wavefront receiving smooth, regular distortions, the transverse intensity distribution becomes inhomogeneous, and the beam broadens out.

The correction of the laser radiation phase, which is a smooth continuous spatial function, can be performed using a conventional adaptive optical system including a wavefront sensor and a wavefront corrector. The wavefront sensor performs the measurement (in other words, reconstruction) of the radiation phase surface; then, on the basis of these data, the wavefront corrector (for example, a reflecting mirror with deformable surface) transforms the phase front in the proper way. If all components of the adaptive optical system are involved in the common circuit with the feedback, then the adaptive system is known as a closed-loop system. The adaptive correction of the wavefront with smooth distortions has a somewhat long history and considerable advances [1, 2, 3, 4, 5, 6].

When a laser beam passes a sufficiently long distance in a turbulent atmosphere, the so-called regime of strong scintillations (intensity fluctuations) is realized. Under such conditions the optical field becomes speckled, lines appear in the space along the beam axis where the intensity vanishes and the surrounding zones of the wavefront attain a helicoidal (screw) shape. If the intensity in an acnode of the transverse plane is zero, then the phase in this point is not defined. In view of its screw form, the phase surface in the vicinity of such point has a break, the height of which is divisible by the wavelength. Since the phase is defined accurate to the addend that is aliquot to 2π, it is formally continuous but under a complete circling on the phase surface around the singular point one cannot reach the starting place. The integration of the phase gradient over some closed contour encircling such singular point results in a circulation not equal to zero, in contrast to the null circulation at the usual smoothed-inhomogeneous regular phase distribution. The indicated properties represent evidence of strong distortions of the wavefront – screw dislocations or optical vortices. The vortical character of the beam is detected with ease in the experiment after the analysis of the picture of its interference with the obliquely incident plane wave: the interference fringes arise or vanish in the centers of screw dislocations forming peculiar "forks".

Scintillations in the atmosphere especially decrease the efficiency of light energy transportation and distort the information carried by a laser beam in issues of astronomy and optical communications. Scintillation effects present special difficulty for adaptive optics, and their correction is one of key trends in the development of state-of-the-art adaptive optical systems.

However it should be noted that the possibility to control the optical vortices (including the means of adaptive optics) presents interest not only for atmospheric optics but for a new optical field, namely, singular optics [7, 8, 9]. The fact is that optical vortices have very promising applications in optical data processing, micro-manipulation, coronagraphy, etc. where any type of management of the singular phase could be required.

This chapter is dedicated to wavefront reconstruction and adaptive phase correction of a vortex laser beam, which is generated in the form of the Laguerre-Gaussian LG_0^1 laser mode. The content of the chapter is as follows. In Section 1 we specify the origin and main properties of optical vortices as well as some their practical applications. Section 2 is dedicated to a short description of the origination of optical vortices in a turbulent atmosphere and correspondent problems of the adaptive optics. In Section 3 some means are given concerning the generation of optical vortices under laboratory conditions, aimed at the formation of a "reference" optical vortex with the maximally predetermined phase surface, and the experimental results of such formation are illustrated. Section 4 is concerned with vortex beam phase surface registration, based on measurements of phase local tilts using a Hartmann-Shack wavefront sensor and a novel reconstruction technique. In Section 5 experimental results of correction of a vortex beam are demonstrated in the conventional closed-loop adaptive optical system including a Hartmann-Shack wavefront sensor and a bimorph deformable adaptive mirror. Conclusions summarize the abovementioned research results.

2. Origin, main properties and practical applications of optical vortices

The singularity of the radiation field phase S is identified by the term "optical vortex", which can appear in the complex function $\exp\{iS\}$ representing a monochromatic light wave [10]. The amplitude of scalar wave field A (and, correspondingly, its intensity $I=|A|^2$) in the point of the vortex location approaches zero. The phase of radiation S changes its value by

$2\pi m$ with the encircling the singularity point clockwise or counter-clockwise, where m is the positive or negative integer number known as the vortex "topological charge". In the centre of the vortex (intensity zero point) the phase remains indefinite. Such an optical singularity is the result of the interference of partial components of the wave field with a phase shift, which is initial or acquired during propagation in an inhomogeneous medium. In 3D space the points with zero intensity form zero lines. On these lines the potentiality of phase field is violated; and the regions of "defective" (singular) phase can be considered as vortex strings like the regions with concentrated vorticity, which are considered in the hydrodynamics of ideal liquid. We are interested in a case where the lines of zero intensity have a predominantly longitudinal direction, i.e. form a longitudinal optical vortex. The equiphase surface in the vicinity of such a line has the appearance of a screw-like (helicoidal) structure, threaded on this line. In the interference pattern of the vortex wave under consideration with any regular wave, the vortices are revealed through the appearance of so-called "forks" (i.e., branching of interference fringes), coinciding with zero points of the intensity.

Investigations of waves with screw wavefront and methods of their generation were reported as early as by Bryngdahl [11]. The theory of waves carrying phase singularities was developed in detail by Nye and Berry [12, 10], prompting a series of publications dealing with the problem (see [13, 14, 7, 8] and the lists of references therein). The term "optical vortex" was introduced in [15]. Along with the term "optical vortex", the phenomenon is also referred to as "wavefront screw dislocation". The latter appeared because of similarities between distorted wavefront and the crystal lattice with defects. The following terms are also used: "topological defects", "phase singularities", "phase cuts", and "branch cuts".

Thus the indication of the existence of an optical vortex in an optical field is the presence of an isolated point {r, z} in a plane, perpendicular to the light propagation axis, in which intensity I (r, z) is approaching zero, phase S (r, z) is indefinite, and integration of the phase gradient $\nabla_\perp S$ field over some closed contour Γ encircling this point results in a circulation not equal to zero:

$$\oint_\Gamma \nabla_\perp S(\boldsymbol{\rho},z)d\boldsymbol{\rho} = 2\pi m, \tag{1}$$

where $d\varrho$ is the element of the contour Γ.

The propagation of slowly-varying complex amplitude of the scalar wave field A in the free space is described by the well-known quasi-optical equation of the parabolic type (see, for example, [16]):

$$\frac{\partial A}{\partial z} - \frac{i}{2k}\frac{\partial^2 A}{\partial r^2} = 0, \tag{2}$$

where $k=2\pi/\lambda$ is wave number, λ is the radiation wavelength, z is the longitudinal coordinate corresponding with the beam propagation axis, $\mathbf{r}=\mathbf{e}_x x+\mathbf{e}_y y$ is the transverse radius-vector. In the process of radiation propagation in the medium vortices appear, travel in space, and disappear (are annihilated).

Laguerre-Gaussian laser beams $LG_n{}^m$ are related to the familiar class of vortex beams and are used most often in experiments with optical vortices [7-9]. They are the eigen-modes of the homogeneous quasi-optical parabolic equation (2) [17], so they do not change their form under the free space propagation and lens transformations. The correspondent solution of equation (2) in cylindrical coordinates (r, φ, z) has the following form:

$$A(r, \varphi; z) = A_0 \frac{w_0}{w} \left(\frac{r}{w}\right)^m \Phi_m(\varphi) L_n{}^m\left(2\frac{r^2}{w^2}\right) \exp\left(-\frac{r^2}{w^2} + i\frac{zr^2}{z_0 w^2} - i(2n + m + 1)\text{arctg}\frac{z}{z_0}\right), \tag{3}$$

The typical transverse size of the beam w in (3) is determined by the relation $w^2 = w_0{}^2 [1+z^2 / z_0{}^2]$ where, in its turn, w_0 is the transverse beam size in the waist and $z_0 = kw_0{}^2 /2$ is the typical waist length. The radial part of distribution (3) includes the generalized Laguerre polynomial

$$L_n{}^m(x) = \frac{e^x}{x^m n!} \frac{d^n}{dx^n}(x^{n+m} e^{-x}). \tag{4}$$

In addition to an item responsible for wavefront curvature and transversely-uniform Gouy phase, the angular factor $\Phi_m(\varphi)$ contributes to the phase part of distribution (3). The angular part of formula (3) is represented in the form of a linear combination of harmonic functions

$$\Phi_m(\varphi) = c_1 \cdot \Omega_m(\varphi) + c_2 \cdot \Omega_{-m}(\varphi), \tag{5}$$

where $\Omega_m(\varphi) = \exp(im\varphi)$, $\varphi = \text{arctg}(y/x)$ is the azimuth angle in the transverse plane. The c_1 and c_2 constants determine the beam character and the presence of singularity in it.

Let's consider two cases of the angular function $\Phi_m(\varphi)$ distribution from (5) at $m>0$. In the first case, when $c_1=1$ and $c_2=0$, we have $\Phi_m(\varphi)=\exp(im\varphi)$. The form of helicoidal phase surface assigned by the $m\varphi$ function (we do not take into account the wavefront curvature in (3) determining the beam broadening or narrowing only) in the vortex Laguerre-Gaussian laser beam is presented in Figure 1. This phase does not depend on r at the given φ and rises linearly with φ increasing. In the phase surface the spatial break of $m\lambda$ (or $2\pi m$ radians) depth is present. Under the complete circular trip around the optical axis on the phase surface it is impossible to get to the starting point. As it has been commented above, such a shape of the phase factor is what causes the singular, vortex behavior of the beam. The positive or negative sign of m determines right or left curling of the phase helix.

On the optical axis, in the vortex center, the intensity is zero, resulting from the behavior of the radial dependence of (3) and generalized Laguerre polynomial (4). The beam intensity distribution in the transverse plane, as it is seen from (3), is axially-symmetrical (modulus of A depends on r only) and visually represents the system of concentric rings. In the simplest case at $n=0$, $m=1$ (LG_0^1 mode) the intensity distribution has a doughnut-like form that is shown in Figure 2. Figure 2 also presents a picture of the interference of the given vortex laser beam with the obliquely incident monochromatic plane wave. In the picture, fringe branching is observed in the beam center with the "fork" formation (fringe birth) typical for screw dislocation [7-9]. At the arbitrary m number the quantity of fringes that are born corresponds with this number, the double, triple, and other "forks" are formed. The presence of "forks" in the interference patterns of such kind is the standard evidence of the vortex nature of the beam in the experiment.

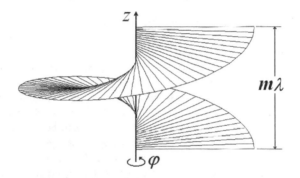

Figure 1. Phase surface shape of a Laguerre-Gaussian beam carrying an optical vortex.

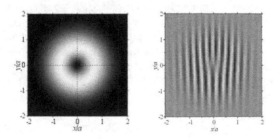

Figure 2. The intensity distribution in the Laguerre-Gaussian laser beam LG_n^m and the picture of its interference with an obliquely incident plane wave at $n=0$, $m=1$, $\Phi_m=\exp(i\varphi)$.

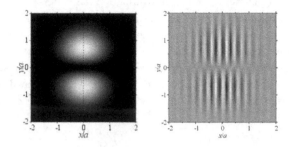

Figure 3. The intensity distribution in the Laguerre-Gaussian laser beam $LG_n{}^m$ and the picture of its interference with an obliquely incident plane wave at $n=0$, $m=1$, $\Phi_m=\sin(i\varphi)$.

Let's consider one more case when $c_1=-c_2=1/2i$ in (5), then $\Phi_m(\varphi)=\sin(m\varphi)$. Under such conditions, according to (3) the intensity distribution has no axial symmetry depending on φ. In Figure 3 the beam intensity distribution in the transverse plane at $\Phi_m=\sin\varphi$ for $n=0$, $m=1$ is shown. The phase portrait of the beam is also given in Figure 3. The half-period shift of interference fringes in the lower half plane as compared to the upper one demonstrates the phase asymmetry with respect to the x axis. The fringe numbers in the lower and upper half planes are the same, i.e. the fringe birth does not take place. It is seen that the edge rather than screw phase dislocation occurs here since the phase distribution is step-like with a break of π (instead of 2π!) radians. The given example of the Laguerre-Gaussian beam demonstrates that at $\Phi_m=\sin\varphi$ the beam is not the vortical in nature. There is no singular point in the beam transverse section but there is a particular line ($y=0$) where the intensity is zero. At $m>1$ in the beam there are m such lines passing through the optical axis and dividing the transverse beam section by $2m$ equal sectors. The radiation phase in each sector is uniform and differs in π in the neighboring sectors.

Light beams with optical vortices currently attract considerable attention. This attention is encouraged by the extraordinary properties of such beams and by the important manifestations of these properties in many applications of science and technology.

It is known for a long time that light with circular polarization possesses an orbital moment. For the single photon its quantity equals $\pm\hbar$, where $\hbar$ is the Planck constant. However only relatively recently it was shown [18, 19] that light can have an orbital moment irrespective of its polarization state if its azimuth phase dependence is of the form $S=m\varphi$ where φ is azimuth coordinate in the transverse cross section of the beam and m is the positive or negative integer. The authors [18] supposed that the moment of each photon is defined by the formula $L=m\hbar$. As this phase dependence is the characteristic feature of the helical wavefront (the form of which is presented in Figure 1) so the beam carrying the optical vortex and possessing such phase front has to own the non-zero orbital moment $m\hbar$ per photon. The quantity of m is defined by the topological charge of optical vortex.

The concept of orbital moment is not new. It is well known that multipole quantum jumps can results in the emission of radiation with orbital moment. However, such processes are infrequent and correspond to some forbidden atomic and molecular transitions. However, generating the beam carrying the optical vortex, one can readily obtain the light radiation beam with quantum orbital moment. Such beams can be used in investigations of all kinds of polarized light. For example, the photon analogy of spin-orbital interaction of electrons can be studied and in general it is possible to organize the search for new optical interactions. As the m-factor can acquire arbitrary values, any part of the beam (even one photon) can carry an unlimited amount of information coded in the topological charge. Thus the density of information in a channel where coding is realized with the use of orbital moment could be as high as compared with a channel with coding of the spin states of a photon. Because only two circular polarization states of the photon are possible, one photon can transmit only one bit of information. Presently, optical vortices have generated a great deal of interest in optical data processing technologies, namely, the coding/decoding in optical communication links in free space [20, 21], optical data processing [22], optical interconnects [23], and quantum optics information processing [24].

The next practical application of optical vortices is optical micromanipulations and construction of so called optical traps, i.e. areas where the small (a few micrometers) particles can be locked in [25, 26]. Progress in the development of such traps allows the capture of particles of low and large refraction indexes [27]. Presently, this direction of research finds further continuation [28, 29, 30].

It is also possible to use optical vortices to register objects with small luminosity located near a bright companion. Shadowing the bright object by a singular phase screen results in the formation of a window, in which the dim object is seen. The optical vortex filtration of such a kind was proposed in [31]. Using this method the companion located at 0.19 arcsec near the object was theoretically differentiated with intensity of radiation 2×10^5 times greater [32]. The possibility to use this method to detect planets orbiting bright stars was also illustrated by astronomers [33, 34]. Vortex coronagraphy is now undergoing further development [35, 36]. There are a number of examples of non-astronomical applications [37, 38].

It was proposed to use optical vortices to improve optical measurements and increase the fidelity of optical testing [39, 40], for investigations in high-resolution fluorescence microscopy [41], optical lithography [42, 43], quantum entanglement [44, 45, 46], Bose–Einstein condensates [47].

Optical vortices show interesting properties in nonlinear optics [48]. For example, in [49, 50, 51] it was predicted that the phase conjugation at SBS of vortex beams is impossible due to the failure of selection of the conjugated mode. For a rather wide class of the vortex laser beams a novel and interesting phenomenon takes place which can be called the phase transformation at SBS. In essence there is only one Stokes mode, the amplification coefficient of which is maximal and higher than that of the conjugated mode. In other words, the non-conjugated mode is selected of in the Stokes beam. The principal Gaussian mode, which is orthogonal to the laser vortex mode, is an example of such an exceptional Stokes mode. The cause of this phenomenon is in the specific radial and azimuth distribution of the vortex laser beam. It is interesting that the hypersound vortices are formed in the SBS medium in accordance with the law of topologic charge conservation. The predicted effects have been completely confirmed experimentally [52, 53, 54].

3. Optical vortices in turbulent atmosphere and the problem of adaptive correction

In early investigations [12] it was shown that the presence of optical vortices is a distinctive property of the so called speckled fields, which form when the laser beam propagates in the scattering media. Experimental evidence of the existence of screw dislocations in the laser beam, passed through a random phase plate, were obtained in [55, 56, 57] where topological limitations were also noted of adaptive control of the laser beams propagating in inhomogeneous media.

Turbulent atmosphere can be represented as the consequence of random phase screens. Under propagation in the turbulent atmosphere the regular optical field acquires rising aberrations. These aberrations manifest themselves in the broadening and random wandering of laser beams; the intensity distribution becomes non-regular and the wavefront deviates from initially set surface. These deformations of the wavefront can be corrected using adaptive optics. To this end, effective sensors and correctors of wavefront were designed [1-6]. The problem becomes more complicated when the laser beam passes a relatively long distance in a weak turbulent medium or if the turbulence becomes too strong. In this case optical vortices develop in the beam; the shape of the wavefront changes qualitatively and singularities appear.

The influence of the scintillation effects are determined (see, for example, [2, 4]) by the closeness to unity of the Rytov variance

$$\sigma_\chi^2 \approx 0.56k^{7/6} \int_0^L C_n^2(z) z^{5/6} (z/L)^{5/6} dz, \tag{6}$$

where $C_n^2(z)$ describes the dependence of structure constant of the refractive index fluctuations over the propagation path and L is the path length. The regime of strong scintillations is not realized when $\sigma_\chi^2 \ll 1$.

Figure 4 demonstrates the results of numerical simulation of propagation of a Gaussian laser beam ($\lambda = 1$ μ) in turbulent atmosphere in a model case of invariable structure constant C_n^2 $= 10^{-14}$ cm$^{-2/3}$ for the distance of 1 km when the regime of strong scintillations is realized according to (6). The steady-state equation for the slowly-varying complex field amplitude A differs from the equation (2) by the presence of the inhomogeneous term:

$$\frac{\partial A}{\partial z} - \frac{i}{2k}\frac{\partial^2 A}{\partial r^2} + \frac{ik}{2}(\tilde{\varepsilon}-1)A=0, \tag{7}$$

where $\tilde{\varepsilon}$ is the fluctuating dielectric permittivity of the turbulent atmosphere. In numerical simulations we use the finite-difference algorithm of numerical solving of parabolic equation (7) described in [58]. It is characterized by an accuracy, which considerably exceeds the accuracy of the widespread spectral methods [59]. The amplitude error of an elementary harmonic solution of the homogeneous equation is equal to zero, whereas the phase error is significantly reduced and proportional to the transverse integration step to the power of six. To take into account the inhomogeneous term of the equation, the splitting by physical processes is employed. The effect of randomly inhomogeneous distribution of dielectric permittivity is allowed for using a model of random phase screens, which is commonly used in calculations of radiation propagation in optically inhomogeneous stochastic media [60]. The spatial spectrum of dielectric permittivity fluctuations is described taking into account the Tatarsky and von Karman modifications of the Kolmogorov model [61].

The fragment of speckled distribution of optical field intensity after the propagation is shown in Figure 4. Dark spots are seen where the intensity vanishes. As it has been noted before, the presence of optical vortices in the beam is easily detected, based on the picture of its interference with an obliquely incident plane wave. The correspondent picture is shown in Figure 4 as well. In the centers of screw dislocations the fringe branching is observed, i.e. the birth or disappearance of the fringes takes place with formation of typical "forks" in the interferogram (compare with Figure 2). There are also zones of edge dislocations (compare with Figure 3). The number, allocation and helicity of the vortices in the beam are random in nature but the vortices are born as well as annihilated in pairs. If the initial beam is regular (vortex-free), then the total topological charge of the vortices in the beam will be equal to zero in each transverse section of the beam along the propagation path in accordance with the conservation law of topological charge (or orbital angular moment) [7-9].

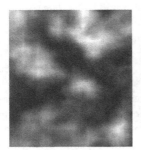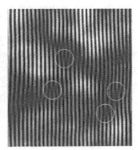

Figure 4. Optical vortices in the laser beam after atmospheric propagation: the speckled intensity distribution and the picture of interference of the beam with the obliquely incident plane wave including "forks" denoted by light circles.

One of the first papers dealing with the appearance of optical vortices in laser beams propagating in randomly inhomogeneous medium was published by Fried and Vaughn in 1992 [62]. They pointed out that the presence of dislocations makes registration of the wavefront more difficult and they considered methods for solving the problem. In 1995 the authors of Ref. [63] encountered this problem in experimental investigations of laser beam propagation in the atmosphere. It was shown that the existence of light vortices is an obstacle for atmospheric adaptive optical systems. After that it was theoretically shown that screw dislocations give rise to errors in the procedure of wavefront registration by the Shack-Hartmann sensor [64, 65]. Due to zero amplitude of the signal in singular points, the information carried by the beam becomes less reliable and the compensation for turbulent aberrations is less effective [66]. Along with [63], the experimental investigation [67] can be taken here as an example where the results of adaptive correction are presented for distortions of beams propagating in the atmosphere.

Since one of the key elements of an adaptive optical system is the wavefront sensor of laser radiation, there is a pressing need to create sensors that are capable of ensuring the required spatial resolution and maximal accuracy of the measurements. In this connection there is necessity need to develop algorithms for measurement of wavefront with screw dislocations, which are sufficiently precise, efficient and economical given the computing resources, and resistant to measurement noises. The traditional methods of wave front measurements [1-6] in the event of the above-mentioned conditions are in fact of no help. The wavefront sensors have been not able to restore the phase under the conditions of strong scintillations [68]. The experimental determination of the location of phase discontinuities itself already generates serious difficulties [69]. In spite of the fact that the construction features of algorithms of wavefront recovery in the presence of screw dislocations were set forth in a number of theoretical papers [68, 69, 70, 71, 72, 73, 74, 75], there were not many published experimental works in this direction. Thus, phase distribution has been investigated in different diffraction orders for a laser beam passed through a specially synthesized hologram, designed for generating higher-order Laguerre-Gaussian modes [76]. An interferometer with high spatial resolution was used to measure transverse phase distribution and localization of phase singularities. The interferometric wavefront sensor was applied also in a high-speed adaptive optical system to compensate phase distortions under conditions of strong scintillations of the coherent radiation in the turbulent atmosphere [77] as well as when modelling the turbulent path under laboratory conditions [78]. In [77, 78] the local phase was measured, without reconstructing the global wavefront that is much less sensitive to the presence of phase residues. The interferometric methods of phase determination are rather complicated and require that several interferograms are obtained at various phase shifts between a plane reference wave and a signal wave. It is noteworthy, however, that in the adaptive optical systems [1-6] the Hartmann-Shack wavefront sensor [79, 80] has a wider application compared with the interferometric sensors including the lateral shearing interferometers [81, 82], the curvature sensor [83, 84, 85], and the pyramidal sensor [86, 87]. The cause of this is just in a simpler and more reliable arrangement and construction of the Hartmann-Shack sensor. However, there have been practically no publications of the results of experimental investigations connected with applications of this sensor for measurements of singular phase distributions.

The problem of a wavefront corrector (adaptive mirror) suitable for controlling a singular phase surface is also topical. In the adaptive optical systems [77, 78] the wavefront correctors were based on the micro-electromechanical system (MEMS) spatial light modulators with the large number of actuators. The results of [77, 78] shown that continuous MEMS mirrors with high dynamic response bandwidth, combined with the interferometric wavefront sensor, can ensure a noticeable correction of scintillation. However, the MEMS mirrors are characterized by low laser damage resistance that can considerably limit applications. The bimorph or pusher-type piezoceramics-based flexible mirrors with the modal response functions of control elements have a much higher laser damage threshold [3-5]. Recently [88] a complicated cascaded imaging adaptive optical system with a number of bimorph piezoceramic mirrors was used to mitigate turbulence effect basing, in particular, on conventional Hartmann-Shack wavefront sensor data. Conventional adaptive compensation was obtained in [88] which proved to be very poor at deep turbulence. The scintillation and vortices may be one of the causes of this.

In the investigations, the results of which are described in this chapter, the development of an algorithm of the Hartmann-Shack reconstruction of vortex wavefront of the laser beam plays a substantial role. The creation of efficient algorithms for the wavefront sensor of vortex beams implies the experiments under modeling conditions when the optical vortices are artificially generated by special laboratory means. Moreover, as long as the matter concerns the creation of a new algorithm of wavefront reconstruction, it is possible to estimate its accuracy only under operation with the beam, the singular phase structure of which is known in detail beforehand. The formation of optical beams with the given configuration of phase singularities and their transformations is one of main trends in the novel advanced optical branch – singular optics [7-9].

Thus, the first stage of the research sees the generation of a vortex laser beam with the given topological charge. In our case the role of this beam is played by the single optical vortex, namely, the Laguerre-Gaussian mode. Further, at the second stage, with the help of the Hartmann-Shack wavefront sensor, the task of registration of the vortex beam phase surface is solved using the new algorithm of singular wavefront reconstruction. Finally, at the third stage, the correction of the singular wavefront is undertaken in a closed-loop adaptive optical system, including the Hartmann-Shack wavefront sensor and the wavefront corrector in the form of a piezoelectric-based bimorph mirror.

4. Generation of optical vortex

As it has been indicated above, to examine the accuracy of the wavefront reconstruction algorithm and its efficiency in the experiment itself a "reference" vortex beam has to be formed with a predetermined phase surface. This is important as, otherwise, it would be impossible to make sure that the algorithm recovers the true phase surface under conditions when robust alternative methods of its reconstruction are missing or unavailable. The Laguerre-Gaussian vortex modes $LG_n{}^m$ can play the role of such "reference" optical vortices.

To create a beam with phase singularities artificially from an initial plane or Gaussian wave, a number of experimental techniques have been elaborated. There are many papers concerning the various aspects of generation of beams with phase singularities (see, for example, [89, 90, 91, 92, 93, 94, 95, 96, 97, 98, 99, 100, 101, 102, 103, 104]). Among other possibilities, we can also refer to several methods for phase singularity creation in the optical beams based on nonlinear effects [105, 106, 107, 108]. The generation of optical vortices is also possible in the waveguides [109, 110, 111]. The adaptive mirrors themselves can be used for the formation of optical vortices [112, 113]. In this chapter, though, we dwell only on a number of ways to generate the vortex beams, which allow one to form close-to-"reference" vortices with well-determined singular phase structure that is necessary for the accuracy analysis of the new algorithm of Hartmann-Shack wavefront reconstruction.

One method for generation of the screw dislocations is by forming the vortex beam immediately inside a laser cavity. The authors of [114] were the first to report that the generation of wavefront vortices is possible using a cw laser source. It was shown in [115] that insertion of a non-axisymmetric transparency into the cavity results in generation of a vortex beam. It was reported in [116] that a pure spiral mode can be obtained by introducing a spiral phase element (SPE) into the laser cavity, which selects the chosen mode. The geometry of the cavity intended for generation of such laser beams from [116] is shown in Figure 5. Here a rear mirror is replaced by a reflecting spiral phase element, which adds the phase change $+2im\varphi$ after reflecting. As a result of reflection, the phase of a spiral mode $-im\varphi$ changes to $+im\varphi$. The cylindrical lens inside the cavity is focused on the output coupler. This lens inverts the helicity of the mode back to the field described by $\exp\{-im\varphi\}$ and ensures the generation of the required spiral beam. The beam at the output passes another cylindrical lens and its distribution becomes the same as inside the cavity. A pinhole in the cavity ensures the generation of a spiral mode of minimal order, i.e., TEM_{01} mode. It should be stressed that the spiral phase element determines the parameters of the spiral beam within the cavity so that by its variation the parameters of the output beam can be controlled.

This method was tested with a linearly polarized CO_2-laser. The reflecting spiral phase element was made of silicon by multilevel etching. It had 32 levels with the entire height of break λ that corresponds to $m=1$. Precision of etching was about 3% and deviation of the surface from the prescribed form was less than 20 nm. The reflecting coefficient of the element was greater than 98%; the diameter and length of the laser tube were 11 mm and 65 cm, respectively. The lens inside the cavity with focal length 12.5 cm was focused in the output concave mirror with a radius of 3 m. An identical lens was placed outside the cavity to collimate the beam. In Figure 6 we demonstrate the stable spiral beam obtained in the experiment [116]. The vortical nature of the beam is proved not by demonstration of the "fork" in the interferogram but by the doughnut-like intensity distribution in the near and the far zone.

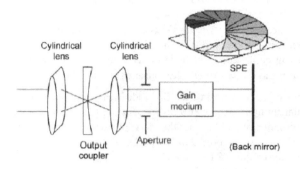

Figure 5. Configuration of a laser cavity intended for generation of spiral beams [116].

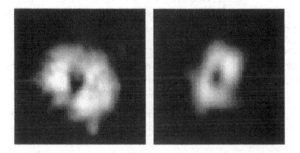

Figure 6. Intensity distribution of vortex beam generated in the laser cavity in near (left) and far (right) zone [116].

A ring cavity with the Dove's prism can also be used to generate vortex beams. It was shown [117] that modes of such a resonator are singular beams.

The next way to generate the optical vortex uses a phase (or a mode) converter. Usually it transforms a Hermit-Gaussian mode, generated in the laser, into a corresponding Laguerre-Gaussian mode. This method was first proposed in [118]. In the experiment the authors used a cylindrical lens, the axis of which was placed at an angle of 45° with respect to the HG_{01} mode for conversion into a LG_0^1 mode. The incident mode appears to the lens as a superposition of HG_{01} mode parallel to the lens axis and a mode perpendicular to the axis. The mode perpendicular to the lens axis passes through a focus, advancing the relative Gouy phase between the two modes of $\pi/2$ as required to form the doughnut mode from uncharged HG_{01} and HG_{10} modes.

In Ref. [91] an expression was derived for an integral transformation of Hermit-Gaussian modes into Laguerre-Gaussian modes in the astigmatic optical system, and it was shown theoretically that passing the beam through the cylindrical lens can perform the conversion. The theory of a $\pi/2$ mode converter and a π mode converter, produced by two cylindrical lenses, was described in more details in [92]. Padgett et al. describe in a tutorial paper [119] how a range of Laguerre-Gaussian modes can be produced using two cylindrical lenses starting from the corresponding Hermit-Gaussian modes, and present the clear examples, showing the intensity and phase distributions obtained. The initial higher order Hermit-Gaussian modes can be produced in the laser with intracavity cross-wires. The authors of [96] used a similar technique with a Nd:YAG laser operating at the 100 mW level.

Even in the absence of the required initial HG_{10} mode in the laser emission, it is easy to produce artificially a similar configuration by introducing a glass plate in a half of the TEM_{00} beam and achieving the necessary π phase shift, and then to apply mode conversion. An example of such doughnut beam creation was reported in [120]. The efficiency of conversion was about 50%. It is possible to use the cylindrical lens mode converter [121] but with production of the initial higher order Hermit-Gaussian mode, by exciting it in an actively stabilized ring cavity, matching in the Gaussian beam from their titanium sapphire laser. The efficiency was up to 40% for the $LG_0{}^1$ mode, and higher order doughnuts could also be obtained easily. The method proposed in [122] is based on the formation of a pseudo HG_{NM} mode, propagating the Gaussian beam of a number of edges of thin glass plates and forming the edge dislocations with the following its astigmatic conversion into a Laguerre-Gaussian mode.

It was reported in [123] that in the event of ideal conversion, the efficiency of Hermit-Gaussian mode transformation into Laguerre-Gaussian mode is about 99.9%. The spherical aberration does not reduce the efficiency factor. Typically cylindrical lenses are not perfect and their defects give rise to several Laguerre-Gaussian modes. The superposition of components can be unstable and this means a dependence of intensity on the longitudinal coordinate. If special means are not employed the precision of lens fabrication is about 5%, in this case the efficiency of beam transformation into Laguerre-Gaussian mode is 95%. Imperfections of 10% result in drop of efficiency down to 80%.

In [124, 54] the formation of the Laguerre-Gaussian $LG_0{}^1$ or $LG_1{}^1$ modes was performed at the output of a pulsed laser-generator of Hermit-Gaussian HG_{01} or HG_{21} modes with the help of a tunable astigmatic $\pi/2$-converter based on the so-called optical quadrupole [125]. It consists of two similar mechano-optical modules, each of which incorporates the positive and negative cylindrical lenses with the same focal length and a positive spherical lens. The mechanical configuration of each module can synchronously turn the incorporated cylindrical lenses in the opposite directions with respect to the optical axis, which ensures its rearrangement. In the initial position the optical forces of cylindrical lenses completely compensate each other and their axes coincide with the main axes of the intensity distribution of the laser. The distance between the modules is fixed so that the spherical lenses in different modules are located at a focal distance from each other and form the optical Fourier transformer.

To study the phase structure of radiation, in [54] use was made of a special interferometer scheme, where the reference beam was produced from a part of the original Laguerre–Gaussian $LG_0{}^1$ or $LG_1{}^1$ mode (see Figure 7). As a result, each of the modes interfered with a similar one, but with a topological charge of the opposite sign (the opposite helicity), i.e., with $LG_0{}^{-1}$ or $LG_1{}^{-1}$ mode. The interference fringe density depended on the thickness of the plane-parallel plate 1 and could be additionally varied by inclining the mirrors 2 (see figure 7).

The peculiarity of the interference of two Laguerre–Gaussian modes, having the opposite helicity of the phase, manifests itself in the branching of a fringe in the middle of the beam and formation of a characteristic "fork" with an additional fringe appearing in the centre, as compared with the case of a vortex mode interfering with a plane reference wave (see Figure 2). Such branching of fringes indicates the vortex nature of the investigated beam, while the absence of branching is a manifestation of the regular character of the beam phase surface.

Figure 8 displays the experimental distributions of intensity of the laser mode LG_0^1 in far field and its picture of interference with an obliquely incident wave in the form of LG_0^{-1} mode.

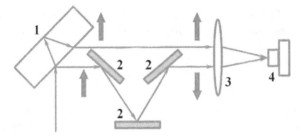

Figure 7. The optical scheme for registration of the phase portrait of a laser beam [54]: 1 – dividing parallel-sided plate, 2 – mirrors, 3 – lens, 4 – CCD camera.

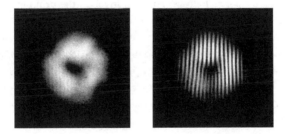

Figure 8. The experimental distribution of intensity and phase portrait of the laser mode LG_0^1 obtained at usage of a phase converter [54].

The invention of a branched hologram [89, 93] uncovered a relatively easy way to produce beams with optical vortices from an ordinary wave by using its diffraction on the amplitude diffraction grating. The idea of singular beam formation is based on the holographic principle: a readout beam restores the wave, which has participated in the hologram recording. Instead of writing a hologram with two actual optical waves, it is sufficient to calculate the interference pattern numerically and, for example, print the picture in black-and-white or grey scale. The amplitude grating after transverse scaling can, when illuminated by a regular wave, reproduce singular beams in diffraction orders.

Using the description of the singular wave amplitude (2), one can easily calculate the pattern of interference of such wave with a coherent plane wave tilted by the angle γ with respect to the z axis. The calculated interference pattern depends on the angle γ between the interfering waves and corresponds to two well-known holographic schemes: on-axis [126] when $\gamma=0$ and off-axis [127] holograms. The spiral hologram (or spiral zone plate) realized under the on-axis scheme suffers from all the disadvantages inherent to the on-axis holograms, namely, the lack of spatial separation of the reconstructed beams from the directly transmitted readout beam. Therefore the on-axis spiral holograms have not found wide application unlike the off-axis computer-generated holograms [128].

Under interference between the plane wave and the optical vortex with unity topologic charge the transmittance of amplitude diffraction grating varies according to

$$T = \left[\frac{1}{2}\left[1 - \cos\left(\left(\frac{2\pi x}{\Lambda}\right) - \arctan\left(\frac{y}{x}\right)\right)\right]\right]^2, \tag{8}$$

where $\Lambda=\lambda/\gamma$ is the grating period. When the basic Gaussian mode passes through the grating and is focused by the lens, the LG_0^1 and LG_0^{-1} modes occur in the far field in the 1st and – 1st orders of diffraction, respectively. The period of the grating should be equal to 100-200 microns to separate the orders of diffraction properly in the actual experiment.

The two simplest ways to fabricate the amplitude diffraction gratings in the form of computer-synthesized holograms are as follows. The first involves the printing of an image onto a transparency utilized in laserjet printers. The second approach consists in photographing an inverted image, printed on a sheet of white paper, onto photo-film. Fragments of images of the gratings with the profile (8) obtained upon usage of the laser transparency with the resolution of 1200 ppi as well as the photo-film are shown in Figure 9 [129, 130, 131]. The usage of the photo-film is more preferable since it gives higher quality of the vortex to be formed and greater power conversion coefficient into the required diffraction order.

Figure 9. Magnified fragment of the amplitude grating in the experiment with laser transparency (left) and photo-film (right).

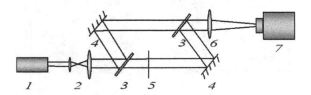

Figure 10. The set-up scheme for formation of the optical vortex:1 – He-Ne laser; 2 – collimator; 3 – optical plane plate, 4 – reflecting plane mirror; 5 – amplitude grating forming the optical vortex; 6 – lens; 7 – CCD camera.

The experimental set-up scheme for formation of the optical vortex with the help of computer-synthesized amplitude grating is shown in Figure 10. The experimental set-up consists of a system for forming the collimated laser beam (λ=0.633 μ), a Mach-Zehnder interferometer and a registration system of the far field beam intensity and interference pattern. The system for forming the collimated beam includes the He-Ne laser 1 and a collimator 2 consisting of two lenses forming the Gaussian beam with a plane wave front. The Mach-Zehnder interferometer consists of two plane plates 3 and two mirrors 4. The computer-synthesized amplitude diffraction grating 5 is inserted into one of the arms of the interferometer. The lens 6 focuses the radiation onto the screen of CCD-camera 7. When blocking or admitting the reference beam from the second arm of the interferometer, the CCD camera registers the far field intensity of the vortex beam or its interference pattern with the reference beam, respectively. Varying the angle between both these beams allows one to vary the interference fringe density.

After passing the beam through the optical scheme, the central peak (0-th diffraction order) is formed in the far-field zone. It concerns the non-scattered component of the beam that has passed through the grating. Less intensive two doughnut-shaped lateral peaks are formed symmetrically from the central peak. The lateral peaks represent the optical vortices and have a topological charge equal to the value and opposite to the sign. In the 1st order of diffraction there is only 16.7 % of energy penetrating the grating in an ideal scenario. In the experiment this part of energy is equal to about 10% owing to the imperfect structure of the grating and its incomplete transmittance.

For registration of the optical vortex it is necessary to cut off the unnecessary diffraction orders. The pictures of doughnut-like intensity distribution of the optical vortex (the lateral peak) in far field and its interference pattern are shown in Figure 11. The rigorous proof of that the obtained lateral peaks bear the optical vortices is the availability of typical "fork" in the interference pattern. The formed vortex in Figure 11, as the vortices in Figures 6 and 8, is rather different from the ideal LG_0^1 mode that is seen from their comparison with Figure 2.

Figure 11. The intensity of the lateral diffraction order in the far field and interference pattern with the plane wave in the case of computer-synthesized amplitude diffraction grating.

Phase transparencies can be used to generate optical vortices. Application of the phase modulator results in phase changes and, after that, in amplitude changes with deep intensity modulation and the advent of zeros. In [132] the optical schematic was described, in which the wave carrying the optical vortex is recorded on thick film (Bregg's hologram) that is used to reproduce the vortex beam. The diffraction efficiency in this schematic is about 99%. A relatively thin transparency with thickness varied gradually in one of the half planes is used in the other method [133]. The efficiency of this method is greater than 90%. A similar method was proposed in [134], but a dielectric wedge was used as the phase modulator. In general, a chain of several vortices is formed as the product of this process. The shape deformation of each vortex depends on the wedge angle and on the diameter of the beam waist on the wedge surface. Varying the waist radius, one can obtain the required number of vortices (even a single vortex).

One more method of the optical vortex generation was proposed in [95]. In this method a phase transparency is used, which immediately adds the artificial vortex component into the phase profile. One such phase modulator is a transparent plate, one surface of which has a helical profile, repeating the singular phase distribution. To obtain Laguerre-Gaussian mode the depth of break onto the surface should be equal to $m\lambda/(n_1 - n_2)$, where n_1 is the plate index of refraction, and n_2 is the medium index of refraction. If these conditions are met, the optical vortex appears in the far field. The main difficulty of this method lies in the problems of fabricating such a transparency. A special mask is used in the manufacture of such plates, which is made negative relative to the spiral phase plate to be formed [135]. The mask is made of brass and checked by the control interferometric system of high precision. After fabrication the mask is filled with a polymer substance and covered by glass. The spiral phase plate is formed on the glass as a result of polymerization. The transfer coefficient of such a plate is 0.98. Publications have appeared recently concerning the generation of phase transparencies using liquid crystals [136, 137].

We note that the manufacture of phase modulators is a special branch of optics called kinoform optics. At the heart of this branch lays the possibility to realize the phase control of radiation by a step-like change of the thickness or the refraction index of some structures [138]. Light weight, small size, and low cost are the most attractive features of kinoform phase elements, when compared with lenses, prisms, mirrors, and other optical devices. The kinoforms can be described as optical elements performing phase modulation with a depth not greater than the wavelength of light. This aim is realized by jumps of the optical path length not less than the even number of half wavelengths. These jumps form the lines dividing the kinoform into several zones. In boundaries of each zone the optical path length can be constant (there are two levels of binary phase elements), they can change discretely (n-level phase elements), or they can change more continuously (in an ideal phase element n approaches infinity). With an increase of n the phase efficiency of the element increases as well as its ability to control light properly. The application of kinoforms facilitates a reduction in the number of optical elements in the system by combining optical properties of several elements into one kinoform. Thus, these optical methods offer broad potential for anyone who wants to obtain beams with desired properties and to generate beams with optical vortices.

The fabrication of spiral (or helicoidal) phase plates techniques has progressed in recent years [139, 140, 141]. We will describe the generation of a doughnut Laguerre-Gaussian LG_0^1 mode with the help of a spiral phase plate [129, 130, 142, 143, 144] manufactured with etching of the fused quartz substrate using kinoform technology. Quartz displays a high damage threshold at λ=0.3-1.3 μ, high uniformity of chemical composition and refractive index n that minimizes the laser beam distortions on passing.

The fabrication of a kinoform spiral phase plate of fused quartz is performed as follows [142]. A quartz plate, 3 cm in diameter and 3 mm in thickness, is taken as the substrate. Both surfaces of the substrate are mechanically polished with a nanodiamond suspension up to the flatness better than λ/30. Special precautions are made to avoid the formation of surface damage layer that may destabilize subsequent etching. A multi-level stepped microrelief, imitating the continuous helicoidal profile, is fabricated using precise sequential etchings of the surface through a photoresist mask in a mixture based on hydrofluoric acid. At every stage, a level pattern is formed in the photoresist layer using a method of deep UV photolithography. The temperature during etching is stabilized with an accuracy of ±0.1°C. The 16- and 32-level spiral phase plates at m=1 and 2 have been fabricated. As the calculations show, such stepped plates and an ideal plate with an exactly helicoidal surface give practically the same optical vortices in the far field. As contrasted to the examples of spiral phase plates [140], the plates [142] have a very high laser damage threshold and a working diameter of 2 cm that is larger by an order of magnitude. The high laser damage resistance of such plates allows their use in experiments with powerful laser beams [52-54]. The general view of the spiral phase plate is shown in Figure 12.

The 3D image of the central part of a 32-level spiral phase plate designed for λ=0.633 μ, m=1 is shown in Figure 13. As it is seen from Figure 13, upon motion around the plate axis along a circle (perpendicular to the propeller bosses), the change of the etched profile altitude is linear with a rather high accuracy. The total break height of the microrelief on the plate surface of 1317.5 nm agrees with the calculated value $\lambda/(n-1)$=1339.6 nm, where n is the substrate refraction index, with an accuracy about of one and a half percent. As the measurements show, the roughness of the etched and non-etched surfaces (including the deepest one) is approximately the same. The roughness rms of each step surface equals 1-1.5 nm, amounting to 2-3% of the height of one step of 43.2 nm.

Figure 12. The photo-image of the 32-level spiral phase plate.

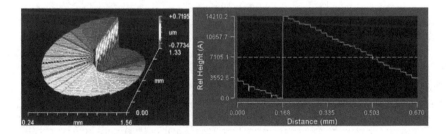

Figure 13. The image of surface in the near-axis region of the 32-level phase plate designed for λ=0.633 μ and its profile shape under motion along the circular line.

It should be noted that a laser beam in the form of a principal Gaussian mode with a plane wavefront that passes through a spiral phase plate maximally resembles the $LG_0{}^1$ mode in the focal plane of a lens (in far field). The part of this mode in the beam exceeds more than 90%; the residual energy is confined in general in the higher Laguerre-Gaussian modes $LG_n{}^1$. It should be noted that the proper intensity modulation of the beam incident to the plate can additionally enhance the portion of the $LG_0{}^1$ mode.

To generate a vortex beam with the help of a spiral phase plate, the experimental setup shown in Figure 10 is used. The spiral phase plate is installed into the scheme instead of the amplitude diffraction grating. In this case a vortex is formed in the 0th diffraction order in far field. Figure 14 demonstrates the experimental distributions of laser intensity in the far field and the pattern of interference of this beam with a obliquely reference plane wave. It is seen that the beam intensity distribution has a true doughnut-like shape. The wavefront singularity appears, as before, by fringe branching in the beam center with the forming of a "fork" typical for screw dislocation with unity topological charge.

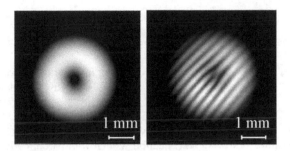

Figure 14. Experimental distribution of intensity of a vortex beam in far field and its interference pattern with obliquely incident reference plane wave in an experiment with a kinoform spiral phase plate.

The experimental data are in good agreement with the results of numerical simulation of the optical system, taking into account the stepped structure of spiral phase plate. The results barely differ from the distribution shown in Figure 2. It should be noted that the vortex quality (similarity to LG_0^1 mode) is very good, caused by the high surface quality of the spiral phase plate throughout its area. This circumstance gives us grounds to believe that the vortex wave front to be reconstructed by the Hartmann-Shack sensor has to be close to the ideal LG_0^1 wave front.

5. Wavefront sensing of optical vortex

The problem of phase reconstruction using the Shack-Hartmann technique was successfully solved for optical fields with smooth wavefronts [145, 146, 147]. In the simplest case, to obtain the phase $S(r)$ using the results of measurement of phase gradient projection $\nabla_\perp S_m(\mathbf{r})$ on the transverse plane it is possible to employ the numeric integration of the gradient over a contour Γ:

$$S(r) = S_0(r) + \int_\Gamma \nabla_\perp S_m(\rho)d\rho, \tag{9}$$

where r={x, y}. Since for the ordinary wave fields the phase distribution is the potential function, the values of $S(r)$ do not depend formally on the configuration of the integration path. In the actual experiments, however, some errors are always present, so the potentiality of phase is violated and the results of phase reconstruction depend on the integration path [147]. To reduce the noise influence on the results it was proposed to consider the phase reconstruction as the minimization of a certain functional. The most commonly used functional is the criterion corresponding to the minimum of the weighted square of residual error of gradient of the reconstructed phase $\nabla_\perp S(\mathbf{r})$ and phase gradients $\nabla_\perp S_m(\mathbf{r})$, obtained in the measurements:

$$\int_D (W(r) \cdot (\nabla_\perp S(r) - \nabla_\perp S_m(r))^2 dr \to \min, \tag{10}$$

where $\mathbf{W}(\mathbf{r})=\{W_x(x, y), W_n(x, y)\}$ is the vector weighting function introduced to account for the reliability of $\nabla_\perp S_m(\mathbf{r})$ measurements. This method is known as the least mean square phase reconstruction.

The approaches to the solution of variation problem (10) are well known [146, 147] and actually mean the solution of the Poisson equation written with partial derivatives. Allowing for the weighting function W(r), it acquires the following form:

$$W_x(x, y)\left(\frac{\partial S}{\partial x} - \frac{\partial S_m}{\partial x}\right) + W_y(x, y)\left(\frac{\partial S}{\partial y} - \frac{\partial S_m}{\partial y}\right) = 0, \tag{11}$$

where $\partial S/\partial x$ and $\partial S/\partial y$ are gradients of reconstructed phase, $\partial S_m/\partial x$ and $\partial S_m/\partial y$ are measured gradients of the phase.

There are a wide variety of methods [145, 146, 147, 148] which can be used to solve the discrete variants of equation (11). For example, one can use the representation of (11) as a system of algebraic equations, the fast Fourier transform, or the Gauss-Zeidel iteration method applied to the multi-grid algorithm. This group of methods is equally well adopted for the application of centroid coordinates measured by the Shack-Hartmann sensor as input data:

$$\{\nabla_\perp S_m(r, z)\}_I = \frac{\iint V(r - r_0) I(r_0) \nabla_\perp S(r_0) dr_0}{\iint V(r - r_0) I(r_0) dr_0}, \tag{12}$$

where the integration is performed over the square of the subaperture, V is a subaperture function, and $I(\mathbf{r}_0)$ is intensity of the input beam.

The sensing of wavefront with screw phase dislocations by the least mean square method is not agreeable. With this technique (along with other methods based on the assumption that phase surface is a continuous function of coordinates) it is possible to reconstruct only a fraction of the entire phase function. As it turned out [68, 149], the differential properties of the vector field of phase gradients help to find some similarity between this field and the field of potential flow of a liquid penetrated by vortex strings. It is also possible to represent this vector field as a sum of potential and solenoid components:

$$\nabla_{\perp} S(\mathbf{r}) = \nabla_{\perp} S_p(\mathbf{r}) + \nabla_{\perp} S_c(\mathbf{r}) \tag{13}$$

where $\nabla_{\perp} S_p$ is the gradient of potential phase component and $\nabla_{\perp} S_c$ is the gradient of vortex (solenoidal) component. By using only the ordinary methods of phase reconstruction it is possible to reproduce just the part of phase distribution that corresponds to potential component in (13).

However, if the quantity $\nabla_{\perp} S_c$ is considered as a rotor of vector potential $\mathbf{H}$, namely, $\nabla_{\perp} S_c = \nabla \times \mathbf{H}$, which is dependent only on the coordinates of optical vortices [68] then potential phase component in (13) can be found by the least mean square method [68, 147]. By means of a novel "hydrodynamic" approach to the properties of the vector filed of phase gradients a new group of methods was formed [72, 150, 151], employing the discovered coordinates of dislocations and reconstructing the potential phase component with the least mean square method. Within another technique [74, 152] reproduction of the scalar potential is also based on the least square method but the vortex component is calculated with Eq. (9) via a consistent rotor of vector potential. The method of matching the vortex component was proposed in [153] and is based on the following equation:

$$\nabla^2 \left(Rot_{-\pi/2}\left(\nabla_{\perp} S_c\right)\right) = \left(\nabla \times \mathbf{H}\right) \cdot \mathbf{e}_z \tag{14}$$

where $\mathbf{e}_z$ is the unit vector of z axis and $Rot_{-\pi/2}(\nabla_{\perp} S_c)$ is operation of the rotation of each vector on $-\pi/2$ angle. This relation is a Poisson equation which allows one to find components of consistent vectors of vortex phase gradient. Now it is possible to take Eq. (9) and obtain a vortex phase component, assuming that the consistent gradients of vortex phase are measured without errors.

The searching for dislocation located positions, which is required in algorithms of phase reconstruction [72, 150, 151], is a sufficiently difficult problem. Because of the infinite phase gradients in the points of zero intensity, the application of methods based on solution of (13) [74, 152] is also not straightforward. Presently there is no such an algorithm, which guarantees the required fidelity of wavefront reconstruction in the presence of dislocations [64]. However, according to some estimations [154, 155, 156] the accurate detection of vortex coordinates and their topological charges insures the sensing of wavefronts with high precision. Therefore we expect a future improvement in reconstruction algorithms by involving more sophisticated methods into the consideration of gradient fields, insuring more accurate detection of dislocation positions and their topological charges.

Analysis shows that from the point of view of experimental realization, of the considered approaches of wavefront reconstruction the algorithm of D. Fried [74] is one of the best algorithms (with respect to accuracy, effectiveness and resistance to measurement noises) of recovery of phase surface $S(x, y)$ from its measured gradient $\nabla S_{\perp}$ distribution in the presence of optical vortices. Fried's algorithm (*a noise-variance-weighted complex exponential reconstructor*) consists of three parts: reduction or simplification, solving, and reconstruction. The algorithm designed for work in Hadjin geometry reconstructs the phase in the nodes of a quadratic grid with the dimensions $(2^N+1)\times(2^N+1)$, using the phase differences between these nodes. Obviously, to employ the algorithm we need $(2^N+1)\times2^N$ and $2^N\times(2^N+1)$ array of phase differences along x and y axes. The words "*complex exponential*" mean that the phase reconstruction problem is reformulated to a task of recovery of "phasors" u (the complex number with a unity absolute value and an argument that is equal to the phase of optical field) distribution in transverse section of the beam. Here, the analysis and transformation of differential complex vectors (differential phasors) $\Delta_x u \equiv \exp(i\Delta_x\varphi)$, $\Delta_y u \equiv \exp(i\Delta_y\varphi)$, corresponding to phase differences $\Delta_x\varphi$, $\Delta_y\varphi$ between different nodes of the computational grid, are used. The words "*noise-variance-weighted*" mean that the algorithm takes into account the distinctions of measurement variance of individual differential phasors, i.e. the influence ("weight") of differential phasors on the recovery result is inversely proportional to their variance. This feature of Fried's algorithm allows us to apply it to a computational grid of arbitrary dimension, not only to the $(2^N+1)\times(2^N+1)$ grid [74]; to take into account the average statistical inequality of measurement errors of phase gradient in different areas of the beam (for example, on the sub-apertures of the Hartmann-Shack sensor) if the repeated characterization of the same beam is performed; to consider a prior concept of the inequality of measurement errors of phase gradient in these areas if the measurement of the beam characteristics is single.

In Fried's algorithm the differential phasors are unit vectors. The operation of normalization of a complex vector is applied to provide for this requirement. However, the amplitudes of differential phasors and phasors, obtained under reduction and reconstruction, contain information about measurement errors of phase differences in the actual experiment. Based on this reason the algorithm in question has been modified [157, 158, 159]. The modification involves exclusion of the operation of complex vector normalization and allows an increase in algorithm accuracy.

The experimental setup for registration of an optical vortex wavefront consists of a system for formation of collimated laser beam, the Mach-Zehnder interferometer (as in the scheme in Figure 10), and the additionally induced the Hartmann-Shack wavefront sensor [160, 161]. It is shown in Figure 15. The system of formation of collimated beam includes a He-Ne laser 1 ($\lambda=0.633$ мкм) and collimator 2 composed of lenses with focal lengths 5 cm and 160 cm. The collimator forms the reference basic Gaussian beam with a diameter of 1 cm and the plane wave front. The Mach-Zehnder interferometer includes two optical plates 3 and two mirrors 4. The spiral phase plate 5 for formation of the optical vortex is interposed into one of the interferometer arms. The working surface of the phase plate, with a diameter of 2 cm completely covers the beam. After passing through the spiral phase plate the Gaussian beam turns into an optical vortex (the LG_0^1 mode) in the focal plane 8 of the lens 6, i.e. in far field, with high conversion coefficient. Focal plane 8 of the lens 6 with focal length 700 cm is transferred (for purposes of magnification) by an objective 8 in the optically conjugated plane 8'. The wavefront sensor consists of a lenslet array 9 situated in the plane 8'and a CCD camera 10.

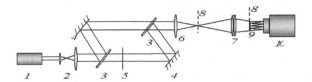

Figure 15. Experimental setup for wavefront sensing of optical vortex in far field: 1 – He-Ne laser; 2 – collimator; 3 – optical plate; 4 – plane mirror; 5 – spiral phase plate; 6 – the lens F=6 m; 7 – the objective; 8 and 8' - focal plane of lens 6 and its optically conjugated plane, respectively; 9 – lenslet array; 10 – CCD camera.

A technical feature of the Hartmann-Shack wavefront sensor used involves the employment of a raster of 8-level diffraction Fresnel lenses as the lenslet array (see Figure 16). The raster is fabricated from fused quartz by kinoform technology, similar to the aforesaid spiral phase plate, with the minimum size of microlens d=0.1 mm and diffraction efficiency up to 90% [162]. The accuracy of etching profile depth is not worse than 2%, the difference of the focal spot size from the theoretical size is ~1%. The spatial resolution of the wavefront sensor and its sensibility depend on the microlens geometry, the number of registered focal spots and their size with respect to the CCD camera pixel size.

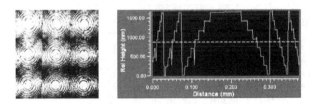

Figure 16. Photo image of a fragment of the lenslet array and image of surface profile of a microlens.

Under the registration of phase front the reference beam in the second arm of the interferometer is blocked. In the beginning the wavefront sensor is calibrated by a reference beam with plane phase front (the spiral phase plate is removed from the scheme). Then the spiral phase plate is inserted, and the picture of focal spots correspondent to singular phase front is registered. From the values of displacement of focal spots from initial positions, the local tilts of wave front on the sub-apertures of lenslet array are determined.

Experiments with a different number of registration spots on the hartmannogram have been carried out [160, 161]. When using a lenslet array with subaperture size d=0.3 mm, focal length f=25 mm and d=0.2 mm, f=15 mm, the picture from 8×8 and 16x16 focal spots on the CCD camera screen has been registered, respectively. The results of experimental measurements of wave front gradients are given in Figure 17 for measurement points 8×8, where the picture of displacements of focal spots in the hartmannogram is shown. The vortex center is situated between the sub-apertures of the array. Displacement of each spot is demonstrated by the arrow (line segment). The arrow origin corresponds to the reference spot position whereas the arrow end corresponds to spot position after the insertion of a spiral phase plate in the experimental scheme. In Figure 17 the results are also shown, which are obtained in calculation and are correspondent to the ideal LG_0^1 mode with high accuracy. It is seen from Figure 17 that the experimental and calculated pictures of spots' displacements agree with each other. Some local data difference is caused by the distinction between the phase and amplitude structure of the beam incident on the phase plate in the calculation and actual experiment, by the inaccurate location of the vortex in the optical axis assumed in calculations and by the inevitable noises of the measurement.

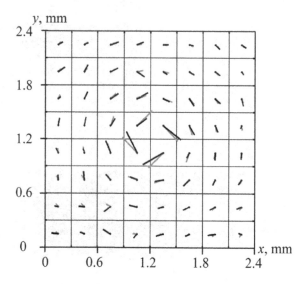

Figure 17. The picture of displacements of focal spots of the hartmannogram in experiment (black arrows) and calculation (grey arrows).

In Ref. [163] the vortex-like structure of displacements of spots in the hartmannogram was registered for the LG_0^1 and higher-order modes. As the primary information, the spot displacements can be used for deriving the Poynting vector skews (in fact, wavefront tilts), as it was made in [163], as well as for wavefront sensing that is more nontrivial. In this chapter we simply consider the reconstruction of singular phase surface by the Hartmann-Shack sensor and describe the realization of this operation with the new reconstruction technique.

In Figure 18 we present the wave front surface of optical vortex reconstructed by the Hartmann-Shack sensor [161, 164] with software incorporating the code of restoration of singular phase surfaces [157-159]. Comparison of experimental data with calculated results shows that the wave front surface is restored by the actual Hartmann Shack wavefront sensor with good quality despite the rather small size of the matrix of wave front tilts (spots in the hartmannogram). The reconstructed wave front has the characteristic spiral form with a break of the surface about 2π. Analysis shows that the accuracy of wave front reconstruction (of course, from the viewpoint of its proximity to the theoretical results) is not worse than $\lambda/20$. The accuracy of recovery of phase surface break increases at the measurement spots of 16x16. For comparison purposes in Figure 18 the result is demonstrated of the vortex wavefront reconstruction with the help of the standard least-squares restoration technique in the Hartmann-Shack sensor. It is seen that the conventional approach obviously fails.

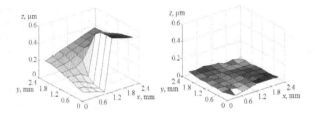

Figure 18. Experimental vortex phase surface reconstructed using modified Fried's (left) and conventional least-squares (right) procedure.

In Figure 19 we show the calculation results [165] of phase front reconstruction of the beam passed through the turbulent atmosphere in the case of $C_n^2=10^{-14}$ cm$^{-2/3}$ after 1 km distance propagation (see Figure 4). The modified Fried's algorithm embedded into the Hartmann-Shack sensor software correctly restores the complicated singular structure of the phase surface.

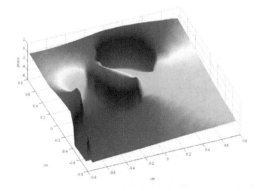

Figure 19. The phase surface fragment of the beam after the turbulent path reconstructed using the modified Fried's algorithm.

6. Phase correction of optical vortex

Next we consider the possibility to transform the wavefronts of the vortex beam by means of the closed-loop adaptive optical system with a wavefront sensor and a flexible deformable wavefront corrector. We can use the bimorph [166] as well as pusher-type [167, 168] piezoceramic-based adaptive mirrors as a wavefront corrector. In the experiments a flexible bimorph mirror [166] and the Hartmann-Shack wavefront sensor with a new reconstruction algorithm [157-159] are employed. An attempt is made to correct the laser beam carrying the optical vortex (namely, the Laguerre-Gaussian LG_0^1 mode), i.e., to remove its singularity. The dynamic effects are not considered, the goal is the estimation of the ability of the bimorph mirror to govern the spatial features of the optical vortex. It is very interesting to determine whether the phase correction leads to full elimination of the singularity [165].

A closed-loop adaptive system intended for performance of the necessary correction of vortex wavefront is shown in Figure 20 [169]. A reference laser beam is formed using a He-Ne laser 1, a collimator 2, and a square pinhole 3, which restricts the beam aperture to a size of 10×10 mm^2. Next the laser beam passes through a 32-level spiral phase plate 5 of a diameter of 2 cm, a fourfold telescope 6 and comes to an adaptive deformable mirror 7. It should be noted that the laser beam with the plane phase front that passes through the spiral phase plate maximally resembles the Laguerre-Gaussian LG_0^1 mode in the far field. In Figure 20 the wavefront corrector is situated in near rather than far field but at a relatively large distance from the spiral phase plate so that the proper vortex structure of the phase distribution is already formed in the wavefront corrector plane.

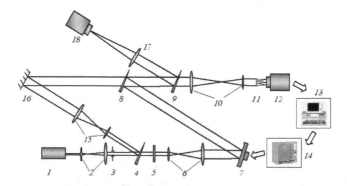

Figure 20. The close-loop adaptive system for optical vortex correction:1 – He-Ne laser; 2 – collimator; 3 – pinhole 10×10 mm; 4, 8, 9 – optical plates; 5 -spiral phase plate; 6, 10, 15 – telescopes; 7 – deformable adaptive mirror; 11 – lenslet array; 12, 18– CCD cameras; 13 –computer; 14 – control unit of adaptive mirror, 16 – plane mirror; 17 – lens.

The wavefront corrector (the bimorph adaptive mirror) 7 [166] is shown in Figure 21. It is composed of a substrate of LK-105 glass with reflecting coating and two foursquare piezoceramic plates, each measuring 45x45 mm and 0.4 mm thick. The first piezoplate is rigidly glued to rear side of the substrate. It is complete, meaning it serves as one electrode, and is intended to compensate for the beam defocusing if need be. The second piezoplate destined to transform the vortex phase surface is glued to the first one. The 5x5=25 electrodes are patterned on the surface of the second piezoplate in the check geometry (close square packing). Each electrode has the shape of a square, with each side measuring 8.5 mm. The full thickness of the adaptive mirror is 4.5 mm. The wavefront corrector is fixed in a metal mounting with a square 45x45 mm window. The surface deformation of the adaptive mirror under the maximal voltage ±300 V applied to any one electrode reaches ±1.5 μm.

25	26	11	12	13
24	10	3	4	14
23	9	2	5	15
22	8	7	6	16
21	20	19	18	17

Figure 21. The deformable bimorph mirror and the scheme of arrangement of control elements on the second piezo-plate.

The radiation beam reflected from the adaptive mirror 7 (see Figure 20) is directed by a plane mirror 8 through a reducing telescope 10 to a Hartmann-Shack sensor including a lenslet array 11 with d=0.2 mm, f=15 mm and a CCD camera 12. At the field size of 3.2 mm there are 16x16 spots in the hartmannogram on the CCD camera screen. The planes of adaptive mirror and lenslet array are optically conjugated so that the wavefront sensor reconstructs in fact the phase surface of the beam just in the corrector plane.

A beam part is derived by a dividing plate 9 to a CCD camera 18 for additional characterization (see Figure 20). In addition, the wavefront corrector 7, plates 4, 8 and rear mirror 16 form a Mach-Zehnder interferometer. On blocking the reference beam from the mirror 16, the CCD cameras 12 and 18 simultaneously register, respectively, the hartmannogram and intensity picture of the beam going from the adaptive mirror. Upon admission of the reference beam from the mirror 16, the CCD camera 18 registers the interference pattern of the beam going from the adaptive mirror with an obliquely incident reference beam. Screen of CCD camera 18 is situated at a focal distance from the lens 17 or in a plane of the adaptive mirror image (like the lenslet array) thus registering the intensity/interferogram of the beam in far or near field, respectively.

The wavefront has no singularity upon removal of the spiral phase plate 5 from the scheme in the Figure 20 and when switching off the wavefront corrector. The reference beam phase surface in the corrector plane is shown in Figure 22a. It is not an ideal plane (PV=0.33 μ) but it is certainly regular. Therefore the picture of diffraction at the square diaphragm 3 (see Figure 20) roughly takes place in far field in Figure 23a.

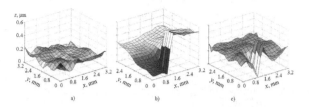

Figure 22. Experimental phase surface in near field: (a) reference beam and beam (b) before and (c) after correction.

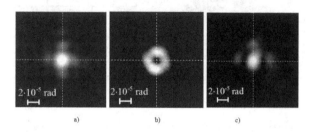

Figure 23. Experimental far field intensity: (a) reference beam and beam (b) before and (c) after correction.

After inserting the spiral phase plate 5 and when switching off the adaptive mirror, the wavefront in near field in Figure 22b acquires the spiral form with λ-break (PV=0.63 μ) so the far-field intensity has a doughnut form (see Figure 23b). Note that the wavefront corrector software in the computer 13 is based on the singular reconstruction technique [157-159]; the conventional least-squares approach fails here. The vortex in far field is the LG_0^1 mode distorted by presence of other modes mainly because of the phase surface imperfection of the reference beam. Note that for the task of adjusting the vortex wavefront sensing technique it was necessary to form a close-to-ideal LG_0^1 mode as a "reference" optical vortex with maximally predetermined phase surface, to determine the sensing algorithm accuracy. Here, for the correction task, it is even more attractive to work with a distorted vortex.

In order to correct the vortex wavefront in the closed loop, the recovered phase surface in Figure 22b is decomposed on the response functions of control elements of the deformable mirror. The response function of a control element is the changing of the shape of the deformable mirror surface upon the energizing of this control element with zero voltages applied to the others actuators. The expansion coefficients on response functions are proportional to voltages to be applied from control unit 14 to appropriate elements of the deformable mirror. When applying control voltages to the adaptive mirror its surface is deformed to reproduce the measured vortex wavefront maximally and thus to obtain a wavefront close to a plane one upon reflection from the corrector. However, each superposition of the response functions of a flexible wavefront corrector is a smooth function, and the corrector is not able to exactly reproduce the phase discontinuity of a depth of 2π. The phase surface after the correction in Figure 22c is close to the reference one (Figure 22a) except for a narrow region at the break line (PV=0.5 μ). As the radiation from this part of the beam is scattered to larger angles and its portion in the beam is relatively small, the far field intensity picture after the correction in Figure 23c is much closer to the reference beam (Figure 23a) rather than the vortex before correction (Figure 23b). Thus, the doughnut-like vortex beam is focused into a beam with a bright axial spot and weaker background that radically increases the Strehl ratio and resolution of the optical system.

The beam interferograms in near field before and after correction are shown in Figure 24. Unlike the former, the latter contains no resolved singularities (at least, under the given fringe density). The vortices, however, may appear under beam propagation from the adaptive mirror plane as it was in the case of combined propagation of the vortex beam with a regular beam [170]. The experimental and calculated (at the reflection of an ideal LG_0^1 mode from the actual deformed adaptive mirror surface) interferograms of a corrected beam in far field are shown in Figure 25. Two off-axis vortices (denoted by light circles) of opposite topological charge are seen here. The first of them is initial vortex shifted from the axis whereas the second arises in the process of beam propagation from the far periphery of the beam (in fact from infinity, according to the terminology of [170]).

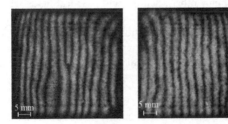

Figure 24. Experimental pattern of interference of the beam with an obliquely incident regular wave in the near field (left) before and (right) after correction.

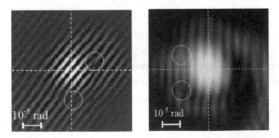

Figure 25. The pattern of interference of the corrected beam with an obliquely incident regular wave in the far field in (left) calculation and (right) experiment.

Thus, the phase surface of the distorted LG_0^1 mode is corrected in the closed-loop adaptive optical system, including the bimorph piezoceramic mirror and the Hartmann-Shack wavefront sensor with the singular reconstruction technique. Experiments demonstrate the ability of the bimorph mirror to correct the optical vortex in a practical sense, namely, to focus the doughnut-like beam into a beam with a bright axial spot that considerably increases the Strehl ratio and optical system resolution. Since the phase break is not reproduced exactly on the flexible corrector surface, the off-axis vortices can appear in far field at the beam periphery.

7. Conclusions

This chapter is dedicated to research of the possibility to control the phase front of a laser beam carrying an optical vortex by means of linear adaptive optics, namely, in the classic closed-loop adaptive system including a Hartmann-Shack wavefront sensor and a deformable mirror. On the one hand, the optical vortices appear randomly under beam propagation in the turbulent atmosphere, and the correction of singular phase front presents a considerable problem for tasks in atmospheric optics, astronomy, and optical communication. On the other hand, the controllable optical vortices have very attractive potential applications in optical data processing and many other scientific and practical fields where the regulation of singular phase is needed. This chapter discusses the main properties and applications of optical vortices, the problem of adaptive correction of singular phase in turbulent atmosphere, the issues of generating the "reference" laser vortex beam, its wavefront sensing and phase correction in the widespread adaptive optical system including a Hartmann-Shack wavefront sensor and a flexible deformable mirror.

The vortex beam is generated with help of a spiral phase plate made of fused quartz by kinoform technology. Provided that the optical quality of the spiral phase plate is good, such a means of vortex formation seems to be more preferable as compared with other considered methods of vortex generation with a well-determined phase surface. As a result, it becomes possible to obtain a singular beam very close to a Laguerre-Gaussian LG_0^1 mode with a well-determined singular phase structure that is necessary for checking the accuracy of subsequent wavefront reconstruction. The developed spiral phase plates are characterized by high laser damage resistance, the good surface profile accuracy and they facilitate formation of a high quality optical vortex.

The vortex phase surface measurement is carried out by a Hartmann-Shack wavefront sensor which is simpler in design and construction, more reliable and more widespread in various fields of adaptive optics when compared with other types of sensors. The commonly accepted Hartmann-Shack wavefront reconstruction is performed on the basis of the least-mean-square approach. This approach works well in the case of continuous phase distributions but is completely unsuitable for singular phase distributions. Therefore a new reconstruction technique has been developed for the reconstruction of singular phase surface, starting from the measured phase gradients. The measured shifts of focal spots in the hartmannogram are in good agreement with the calculation results. Using new software in the Hartmann-Shack sensor, the reconstruction of the "reference" vortex phase surface has been carried out to a high degree of accuracy.

The vortex laser beam (distorted LG_0^1 mode) is corrected in the closed-loop adaptive system including a Hartmann-Shack wavefront sensor with singular reconstruction technique and a flexible bimorph piezoceramic mirror with 5x5 actuators allocated in the check geometry. The mirror has high laser damage resistance meaning it can operate with powerful laser beams. The purpose of the correction is to eliminate the singularity of the beam to the highest degree possible. Experiments have demonstrated the ability of the bimorph mirror to correct the optical vortex in a practical sense. As a result of phase correction, the doughnut-like beam is focused into a beam with a bright axial spot that considerably increases the Strehl ratio and is important for practical applications. However, since the wavefront break cannot be reproduced exactly by a mirror with a flexible surface, the residual off-axis vortices can appear in far field at the beam periphery.

The investigations described above consolidate the actual birth of the experimental field of novel scientific branch – singular adaptive optics.

Author details

S. G. Garanin[1*], F. A. Starikov[1] and Yu. I. Malakhov[2*]

*Address all correspondence to: malakhov@istc.ru

1 Russian Federal Nuclear Center –VNIIEF, Institute of Laser Physics Research, Russia

2 International Science and Technology Center, Russia

References

[1] Vorontsov, M. A., & Shmalgauzen, V. I. (1985). Principles of adaptive optics. *Moscow: Nauka*.

[2] Roggemann, M. C., & Welsh, B. M. (1996). Imaging through turbulence. *Boca Raton, FL: CRC Press*.

[3] Tyson, R. K. (1998). Principles of adaptive optics. *Boston, MA: Academic*.

[4] Hardy, J. W. (1998). Adaptive optics for astronomical telescopes. *New York: Oxford University Press*.

[5] Roddier, F. (1999). Adaptive optics in astronomy. *Cambridge: Cambridge University Press*.

[6] Tyson, R. K. (2000). Introduction to adaptive optics. *Bellingham, WA: SPIE, International Society for Optical Engineering.*

[7] Optical Vortices. (1999). *Vasnetsov M, Staliunas K. (eds.) Horizons in World Physics, 228,* New York: Nova Science.

[8] Soskin, M. S., & Vasnetsov, M. V. (2001). Singular optics. *Wolf E. (ed.) Progress in Optics, V.XLII. Amsterdam: Elsevier, 219-276.*

[9] Bekshaev, A., Soskin, M., & Vasnetsov, M. (2009). Paraxial light beams with angular momentum. *Schulz M. (ed.) Progress in Optics Research. New York: Nova Science Publishers, 1-75.*

[10] Berry, M. (1981). Singularities in waves and rays. *Balian R., Kleman M., Poirier J.-P. (ed.) Physics of Defects. Amsterdam: North-Holland, 453-543.*

[11] Bryngdahl, O. (1973). Radial- and circular-fringe interferograms. *J. Opt. Soc. Am,* 63(9), 1098-1104.

[12] Nye, J. F., & Berry, M. V. (1974). Dislocations in wave trains. *Proc. R. Soc. A,* 336, 165-190.

[13] Rozas, D., Law, C. T., & Swartzlander, G. A. Jr. (1997). Propagation dynamics of optical vortices. *J. Opt. Soc. Am. B,* 14(11), 3054-3065.

[14] Sacks, Z. S., Rozas, D., & Swartzlander, G. A. Jr. (1998). Holographic formation of optical-vortex filaments. *Jr. Opt. Soc. Am. B,* 15(8), 2226.

[15] Coullet, P., Gil, L., & Rocca, F. (1989). Optical vortices. *Optics Commun,* 73(5), 403-408.

[16] Siegman, A. E. (1986). Lasers. *Sausalito, CA: University Science Books.*

[17] Kogelnik, H., & Li, T. (1966). Laser beams and resonators. *Applied Optics,* 5(10), 1550-1567.

[18] Allen, L., Beijersbergen, M. W., Spreeuw, R. J. C., & Woerdman, J. P. (1992). Orbital angular momentum of light and the transformation of Laguerre-Gaussian modes. *Phys Rev. A,* 45(11), 8185-8189.

[19] Allen, L., Padjett, M. J., & Babiker, M. (1999). Orbital angular momentum of light. *Wolf E. (ed.) Progress in Optics,* 34, Amsterdam: Elsevier, 291-370.

[20] Miller, D. A. B. (1998). Spatial channels for communicating with waves between volumes. *Optics Letters,* 23(21), 1645-1647.

[21] Gibson, G., Courtial, J., Padgett, M., Vasnetsov, M., Pas'ko, V., Barnett, S., & Franke-Arnold, S. (2004). Free-space information transfer using light beams carrying orbital angular momentum. *Optics Express,* 12(22), 5448-5456.

[22] Bouchal, Z., & Celechovsky, R. (2004). Mixed vortex states of light as information carriers. *New J. Phys,* 6(1), 131-145.

[23] Scheuer, J., & Orenstein, M. (1999). Optical vortices crystals: spontaneous generation in nonlinear semiconductor microcavities. *Science*, 285(5425), 230-233.

[24] Mandel, L., & Wolf, E. (1995). Optical Coherence and Quantum Optics. *New York: Cambridge University Press*.

[25] Ashkin, A. (1992). Forces of a single-beam gradient laser trap on a dielectric sphere in the ray optics regime. *Biophys. J*, 61(2), 569-582.

[26] Gahagan, K. T., & Swartzlander, G. A. Jr. (1998). Trapping of low-index microparticles in an optical vortex. *J. Opt. Soc. Am. B*, 15(2), 524-534.

[27] Gahagan, K. T., & Swartzlander, G. A. Jr. (1999). Simultaneous trapping of low-index and high-index microparticles observed with an optical-vortex trap. *J. Opt. Soc. Am. B*, 16(4), 533-539.

[28] Curtis, J. E., Koss, B. A., & Grier, D. G. (2002). Dynamic holographic optical tweezers. *Optics Commun*, 207(1-6), 169-175.

[29] Ladavac, K., & Grier, D. G. (2004). Microoptomechanical pump assembled and driven by holo- graphic optical vortex arrays. *Optics Express*, 12(6), 1144-1149.

[30] Daria, V., Rodrigo, P. J., & Glueckstad, J. (2004). Dynamic array of dark optical traps. *Appl. Phys. Lett*, 84(3), 323-325.

[31] Khonina, S. N., Kotlyar, V. V., Shinkaryev, M. V., Soifer, V. A., & Uspleniev, G. V. (1992). *J. Mod. Optics*, 39(5), 1147-1154.

[32] Swartzlander, G. A. Jr. (2001). Peering into darkness with a vortex spatial filter. *Optics Letters*, 26(8), 497-499.

[33] Rouan, D., Riaud, P., Boccaletti, A., Clénet, Y., & Labeyrie, A. (2000). The four-quadrant phase-mask coronagraph. I. Principle. *Pub. Astron. Soc. Pacific*, 112, 1479-1486.

[34] Boccaletti, A., Riaud, P., Baudoz, P., Baudrand, J., Rouan, D., Gratadour, D., Lacombe, F., & Lagrange-M, A. (2004). The four-quadrant phase-mask coronagraph. IV. First light at the very large telescope. *Pub. Astron. Soc. Pacific*, 116, 1061-1071.

[35] Foo, G., Palacios, D. M., & Swartzlander, G. A. Jr. (2005). Optical vortex coronagraph. *Optics Letters*, 30(24), 3308-3310.

[36] Lee, J. H., Foo, G., Johnson, E. G., & Swartzlander, G. A. Jr. (2006). Experimental verification of an optical vortex coronagraph. *Phys. Rev. Lett*, 97(5), 053901-1.

[37] Davis, J. A., Mc Namara, D. E., & Cottrell, D. M. (2000). Image processing with the radial Hilbert transform: theory and experiments. *Optics Letters*, 25(2), 99-101.

[38] Larkin, K. G., Bone, D. J., & Oldfield, M. A. (2001). Natural demodulation of two-dimensional fringe patterns. I. General background of the spiral phase quadrature transform. *J. Opt. Soc. Am. A*, 18(8), 1862-1870.

[39] Masajada, J., Popiołek-Masajada, A., & Wieliczka, D. M. (2002). The interferometric system using optical vortices as phase markers. *Optics Commun*, 207(1), 85-93.

[40] Senthilkumaran, P. (2003). Optical phase singularities in detection of laser beam collimation. *Applied Optics*, 42(31), 6314-6320.

[41] Westphal, V., & Hell, S. W. (2005). Nanoscale Resolution in the Focal Plane of an Optical Microscope. *Phys. Rev. Lett*, 94(14), 143903-1.

[42] Levenson, M. D., Ebihara, T. J., Dai, G., Morikawa, Y., Hayashi, N., & Tan, S. M. (2004). Optical vortex mask via levels. *J. Microlithogr. Microfabr. Microsyst*, 3(2), 293-304.

[43] Menon, R., & Smith, H. I. (2006). Absorbance-modulation optical lithography. *J. Opt. Soc. Am. A*, 23(9), 2290-2294.

[44] Mair, A., Vaziri, A., Weihs, G., & Zeilinger, A. (2001). Entanglement of the orbital angular momentum states of photons. *Nature*, 412(7), 313-316.

[45] Arnaut, H. H., & Barbosa, G. A. (2000). Orbital and intrinsic angular momentum of single photons and entangled pairs of photons generated by parametric down-conversion. *Phys. Rev. Lett*, 85(2), 286-289.

[46] Franke-Arnold, S., Barnett, S. M., Padgett, M. J., & Allen, L. (2002). Two-photon entanglement of orbital angular momentum states. *Phys. Rev. A*, 65(3), 033823.

[47] Abo-Shaeer, J. R., Raman, C., Vogels, J. M., & Ketterle, W. (2001). Observation of vortex lattices in Bose-Einstein condensates. *Science*, 292(5516), 476-479.

[48] Dholakia, K., Simpson, N. B., Padgett, M. J., & Allen, L. (1996). Second-harmonic generation and the orbital angular momentum of light. *Phys. Rev. A*, 54(5), R3742-3745.

[49] Starikov, F. A., & Kochemasov, G. G. (2001). Novel phenomena at stimulated Brillouin scattering of vortex laser beams. *Optics Commun*, 193(1-6), 207-215.

[50] Starikov, F. A., & Kochemasov, G. G. (2001). Investigation of stimulated Brillouin scattering of vortex laser beams. *Proc. SPIE*, 4403-217.

[51] Starikov, F. A. (2007). Stimulated Brillouin scattering of Laguerre-Gaussian laser modes: new phenomena. *Gaponov-Grekhov AV., Nekorkin VI. (ed). Nonlinear waves 2006. N.Novgorod: IAP RAS*, 206-221.

[52] Starikov, F. A., Dolgopolov, Yu. V., Kopalkin, A. V., et al. (2006). About the correction of laser beams with phase front vortex. *J. Phys. IV*, 133, 683-685.

[53] Starikov, F. A., Dolgopolov, Yu. V., Kopalkin, A. V., et al. (2008). New phenomena at stimulated Brillouin scattering of Laguerre-Gaussian laser modes: theory, calculation, and experiments. *Proc. SPIE*, 70090E, 1-11.

[54] Kopalkin, A. V., Bogachev, V. A., Dolgopolov, Yu. V., et al. (2011). Conjugation and transformation of the wave front by stimulated Brillouin scattering of vortex Laguerre-Gaussian laser modes. *Quantum Electronics*, 41(11), 1023-1026.

[55] Baranova, N. B., Zel'dovich, B., Ya., Mamaev. A. V., Pilipetskii, N. V., & Shkunov, V. V. (1981). Dislocations of the wavefront of a speckle-inhomogeneous field (theory and experiment). *Sov. Phys. JETP Lett*, 33(4), 195-199.

[56] Baranova, N. B., Zel'dovich, B. Ya., Mamaev, A. V., Pilipetskii, N. V., & Shkunov, V. V. (1982). Dislocation density on wavefront of a speckle-structure light field. *Sov. Phys. JETP*, 56(5), 983-988.

[57] Baranova, N. B., Mamaev, A. V., Pilipetskii, N. V., Shkunov, V. V., & Zel'dovich, B. Ya. (1983). Wavefront dislocations: topological limitations for adaptive systems with phase conjugation. *J. Opt. Soc. Am. A*, 73(5), 525-528.

[58] Ladagin, V. K. (1985). About the numerical integration of a quasi-optical equation. *Questions of Atomic Science and Technology. Ser. Methods and codes of numerical solution of tasks of mathematical physics* [1], 19-26.

[59] Feit, M. D., & Fleck, J. A. Jr. (1988). Beam nonparaxiality, filament formation, and beam breakup in the self-focusing of optical beams. *J. Opt. Soc. Am. B*, 7(3), 633-640.

[60] Kandidov, V. P. (1996). Monte Carlo method in nonlinear statistical optics. *Physics Usp*, 39(12), 1243-1272.

[61] Goodman, J. W. (2000). Statistical optics. *New York: Wiley*.

[62] Fried, D. L., & Vaughn, J. L. (1992). Branch cuts in the phase function. *Applied Optics*, 31(15), 2865-2882.

[63] Primmerman, A., Pries, R., Humphreys, R. A., Zollars, B. G., Barclay, H. T., & Herrmann, J. (1995). Atmospheric-compensation experiments in strong-scintillation conditions. *Applied Optics*, 34(12), 081-088.

[64] Barchers, J. D., Fried, D. L., & Link, D. J. (2002). Evaluation of the performance of Hartmann sensors in strong scintillation. *Applied Optics*, 41(6), 1012-1021.

[65] Kanev, F. Yu., Lukin, V. P., & Makenova, N. A. (2002). Analysis of adaptive correction efficiency with account of limitations induced by Shack-Hartmann sensor. *Proc. SPIE*, 5026, 190-197.

[66] Ricklin, J. C., & Davidson, F. M. (1998). Atmospheric turbulence effects on a partially coherent Gaussian beam: implication for free-space laser communication. *Applied Optics*, 37(21), 4553-4561.

[67] Levine, M., Martinsen, E. A., Wirth, A., Jankevich, A., Toledo-Quinones, M., Landers, F., & Bruno, Th. L. (1998). Horizontal line-of-sight turbulence over near-ground paths and implication for adaptive optics corrections in laser communications. *Applied Optics*, 37(21), 4553-4561.

[68] Fried, D. L. (1998). Branch point problem in adaptive optics. *J. Opt. Soc. Am. A*, 15(10), 2759-2768.

[69] Le Bigot, E. O., Wild, W. J., & Kibblewhite, E. J. (1998). Reconstructions of discontinuous light phase functions. *Optics Letters*, 23(1), 10-12.

[70] Takijo, H., & Takahashi, T. (1988). Least-squares phase estimation from the phase difference. *J. Opt. Soc. Am. A*, 5(3), 416-425.

[71] Aksenov, V. P., Banakh, V. A., & Tikhomirova, O. V. (1998). Potential and vortex features of optical speckle field and visualization of wave-front singularities. *Applied Optics*, 37(21), 4536-4540.

[72] Arrasmith, W. W. (1999). Branch-point-tolerant least-squares phase reconstructor. *J. Opt. Soc. Am. A*, 16(7), 1864-1872.

[73] Tyler, G. A. (2000). Reconstruction and assessment of the least-squares and slope discrepancy components of the phase. *J. Opt. Soc. Am. A*, 17(10), 1828-1839.

[74] Fried, D. L. (2001). Adaptive optics wave function reconstruction and phase unwrapping when branch points are present. *Optics Commun*, 200(1), 43-72.

[75] Aksenov, V. P., & Tikhomirova, O. V. (2002). Theory of singular-phase reconstruction for an optical speckle field in the turbulent atmosphere. *J. Opt. Soc. Am. A*, 19(2), 345-355.

[76] Rockstuhl, C., Ivanovskyy, A. A., Soskin, M. S., et al. (2004). High-resolution measurement of phase singularities produced by computer-generated holograms. *Optics Commun*, 242(1-3), 163-169.

[77] Baker, K. L., Stappaerts, E. A., Gavel, D., et al. (2004). High-speed horizontal-path atmospheric turbulence correction with a large-actuator-number microelectromechanical system spatial light modulator in an interferometric phase-conjugation engine. *Optics Letters*, 29(15), 1781-1783.

[78] Notaras, J., & Paterson, C. (2007). Demonstration of closed-loop adaptive optics with a point-diffraction interferometer in strong scintillation with optical vortices. *Optics Express*, 15(21), 13745-13756.

[79] Hartmann, J. (1904). Objetivuntersuchungen. *Z. Instrum* [1, 1], 33-97.

[80] Shack, R. B., & Platt, B. C. (1971). Production and use of a lenticular Hartmann screen. *J. Opt. Soc. Am*, 6(5), 656-662.

[81] Hardy, J. W., Lefebvre, J. E., & Koliopoulos, C. L. (1977). Real-time atmospheric compensation. *J. Opt. Soc. Am*, 67(3), 360-369.

[82] Sandler, D. G., Cuellar, L., Lefebvre, M., et al. (1994). Shearing interferometry for laser-guide-star atmospheric correction at large D/r_0. *J. Opt. Soc. Am. A*, 11(2), 858-873.

[83] Roddier, F. (1988). Curvature sensing and compensation: a new concept in adaptive optics. *Applied Optics*, 27(7), 1223-1225.

[84] Rousset, G. (1999). Wave-front sensors. *Roddier F. (ed.) Adaptive optics in astronomy- Cambridge: Cambridge University Press*, 91-130.

[85] Dorn, R. J. (2001). A CCD based curvature wavefront sensor for adaptive optics in astronomy. *Dissertation for the degree of Doctor of Natural Sciences. University of Heidelberg, Germany.*

[86] Ragazzoni, R. (1996). Pupil plane wavefront sensing with an oscillating prism. *Journal of Modern Optics,* 43(2), 289-293.

[87] Ragazzoni, R., Ghedina, A., Baruffolo, A., Marchetti, E., et al. (2000). Testing the pyramid wavefront sensor on the sky. *Proc. SPIE,* 4007, 423-429.

[88] Vorontsov, M., Riker, J., Carhart, G., Rao Gudimetla, V. S., Beresnev, L., Weyrauch, T., & Roberts, L. C. Jr. (2009). Deep turbulence effects compensation experiments with a cascaded adaptive optics system using a 3.63 m telescope. *Applied Optics,* 48(1), A47-57.

[89] Bazhenov, V. Yu., Vasnetsov, M. V., & Soskin, M. S. (1990). Laser beams with screw wavefront dislocations. *Sov. Phys. JETP Lett,* 52(8), 429-431.

[90] Brambilla, M., Battipede, F., Lugiato, L. A., Penna, V., Prati, F., Tamm, C., & Weiss, C. O. (1991). Transverse Laser Patterns. I. Phase singularity crystals. *Phys. Rev. A,* 43(9), 5090-5113.

[91] Abramochkin, E., & Volostnikov, V. (1991). Beam transformations and nontransformed beams. *Optics Commun,* 83(1, 2), 123-135.

[92] Grin', L. E., Korolenko, P. V., & Fedotov, N. N. (1992). About the generation of laser beams with screw wavefront structure. *Optics and Spectroscopy,* 73(5), 1007-1010.

[93] Bazhenov, V. Y.u, Soskin, M. S., & Vasnetsov, M. V. (1992). Screw dislocations in light wavefronts. *J. Mod. Optics,* 39(5), 985-990.

[94] Beijersbergen, M. W., Allen, L., van der Veen, H. E. L. O., & Woerdman, J. P. (1993). Astigmatic laser mode converters and transfer of orbital angular momentum. *Optics Commun,* 96(1-3), 123-132.

[95] Beijersbergen, M. W., Coerwinkel, R. P. C., Kristensen, M., & Woerdman, J. P. (1994). Helical- wavefront laser beams produced with a spiral phase plate. *Optics Commun,* 112(5-6), 321-327.

[96] Dholakia, K., Simpson, N. B., Padgett, M. J., & Allen, L. (1996). Second harmonic generation and the orbital angular momentum of light. *Phys. Rev. A,* 54(5), R3742-R3745.

[97] Oron, R., Danziger, Y., Davidson, N., Friesem, A., & Hasman, E. (1999). Laser mode discrimination with intra-cavity spiral phase elements. *Optics Commun,* 169(1-6), 115.

[98] Wada, A., Miyamoto, Y., Ohtani, T., Nishihara, N., & Takeda, M. (2001). Effects of astigmatic aberration in holographic generation of Laguerre-Gaussian beam. *Proc. SPIE,* 4416, 376-379.

[99] Miyamoto, Y., Masuda, M., Wada, A., & Takeda, M. (2001). Electron-beam lithography fabrication of phase holograms to generate Laguerre-Gaussian beams. *Proc. SPIE,* 3740, 232-235.

[100] Zhang, D. W., & Yuan X., -C. (2002). Optical doughnut for optical tweezers. *Optics Letters,* 27(15), 1351-1353.

[101] Malyutin, A. A. (2004). On a method for obtaining laser beams with a phase singularity. *Quantum Electronics,* 34(3), 255-260.

[102] Izdebskaya, Y., Shvedov, V., & Volyar, A. (2005). Generation of higher-order optical vortices by a dielectric wedge. *Optics Letters,* 30(18), 2472-2474.

[103] Vyas, S., & Senthilkumaran, P. (2007). Interferometric optical vortex array generator. *Applied Optics,* 46(15), 2893-2898.

[104] Kotlyar, V. V., & Kovalev, A. A. (2008). Fraunhofer diffraction of the plane wave by a multilevel (quantized) spiral phase plate. *Optics Letters,* 33(2), 189-191.

[105] Arecchi, F. T., Boccaletti, S., Giacomelli, G., Puccioni, G. P., Ramazza, P. L., & Residori, S. (1992). Patterns, space-time chaos and topological defects in nonlinear optics. *Physica D: Nonlinear Phenomena,* 61(1-4), 25-39.

[106] Indebetouw, G., & Korwan, D. R. (1994). Model of vortices nucleation in a photorefractive phase-conjugate resonator. *J. Mod. Optics,* 41(5), 941-950.

[107] Soskin, M. S., & Vasnetsov, M. V. (1998). Nonlinear singular optics. *Pure and Applied Optics,* 7(2), 301-311.

[108] Berzanskis, A., Matijosius, A., Piskarskas, A., Smilgevicius, V., & Stabinis, A. (1997). Conversion of topological charge of optical vortices in a parametric frequency converter. *Optics Commun,* 140(4-6), 273-276.

[109] Yin, J., Zhu, Y., Wang, W., Wang, Y., & Jhe, W. (1998). Optical potential for atom guidance in a dark hollow beam. *J. Opt. Soc. Am. B,* 15(1), 25-33.

[110] Darsht, B., Ya., Zel'dovich. B., Ya., Kataevskaya. I. V., & Kundikova, N. D. (1995). Formation of a single wavefront dislocation. *Zh. Eksp. Teor. Fiz,* 107(5), 1464-1472.

[111] Fadeeva, T. A., Reshetnikoff, S. A., & Volyar, A. V. (1998). Guided optical vortices and their angular momentum in low-mode fibers. *Proc. SPIE,* 3487, 59-70.

[112] Sobolev, A., Cherezova, T., Samarkin, V., & Kydryashov, A. (2007). Bimorph flexible mirror for vortex beam formation. *Proc. SPIE,* 63462A, 1-8.

[113] Tyson, R. K., Scipioni, M., & Viegas, J. (2008). Generation of an optical vortex with a segmented deformable mirror. *Applied Optics,* 47(33), 6300-6306.

[114] Vaughan, J. M., & Willets, D. V. (1979). Interference properties of a light beam having a helical wave surface. *Optics Commun,* 30(3), 263-270.

[115] Rozanov, N. N. (1993). About the formation of radiation with wavefront dislocations. *Optics and Spectroscopy*, 75(4), 861-867.

[116] Oron, R., Davidson, N., Friesem, A. A., & Hasman, E. (2000). Efficient formation of pure helical laser beams. *Optics Commun*, 182(1-3), 205-208.

[117] Abramochkin, E., Losevsky, N., & Volostnikov, V. (1997). Generation of spiral-type laser beams. *Optics Commun*, 141(1-2), 59-64.

[118] Tamm, C., & Weiss, C. O. (1990). Bistability and optical switching of spatial patterns in laser. *J. Opt. Soc. Am. B*, 7(6), 1034-1038.

[119] Padgett, M., Arlt, J., Simpson, N., & Allen, L. (1996). An experiment to observe the intensity and phase structure of Laguerre-Gaussian laser modes. *Am. J. Phys*, 64(1), 77-82.

[120] Petrov, D. V., Canal, F., & Torner, L. (1997). A simple method to generate optical beams with a screw phase dislocation. *Optics Commun*, 143(4-6), 265.

[121] Snadden, M. J., Bell, A. S., Clarke, R. B. M., Riis, E., & Mc Intyre, D. H. (1997). Doughnut mode magneto-optical trap. *J. Opt. Soc. Am. B*, 14(3), 544-552.

[122] Yoshikawa, Y., & Sasada, H. (2002). Versatile of optical vortices based on paraxial mode expansion. *J. Opt. Soc. Am. A*, 19(10), 2127-2133.

[123] Courtial, J., & Padjett, M. J. (1999). Performance of a cylindrical lens mode converter for producing Laguerre-Gaussian laser modes. *Optics Commun*, 159(1-3), 13-18.

[124] Bagdasarov, V. Kh., Garnov, S. V., Denisov, N. N., Malyutin, A. A., Dolgopolov, Yu. V., Kopalkin, A. V., & Starikov, F. A. (2009). Laser system emitting 100 mJ in Laguerre-Gaussian modes. *Quantum Electronics*, 39(9), 785-788.

[125] Malyutin, A. A. (2006). Tunable astigmatic $\pi/2$ converter of laser modes with a fixed distance between input and output planes. *Quantum Electronics*, 36(1), 76-78.

[126] Gabor, D. (1948). A new microscopic principle. *Nature*, 161(4098), 777-778.

[127] Leith, E., & Upatnieks, J. (1961). New technique in wavefront reconstruction. *J. Opt. Soc. Am*, 51(11), 1469-1473.

[128] Heckenberg, N. R., Mc Duff, R., Smith, CP, & White, A. G. (1992). Generation of optical phase singularities by computer-generated holograms. *Optics Letters*, 17(3), 221-223.

[129] Starikov, F. A., Atuchin, V. V., Dolgopolov, Yu. V., et al. (2004). Generation of optical vortex for an adaptive optical system for phase correction of laser beams with wave front dislocations. *Proc. SPIE*, 5572, 400-408.

[130] Starikov, F. A., Atuchin, V. V., Dolgopolov, Yu. V., et al. (2005). Development of an adaptive optical system for phase correction of laser beams with wave front dislocations: generation of an optical vortex. *Proc. SPIE*, 5777, 784-787.

[131] Starikov, F. A., & Kochemasov, G. G. (2005). ISTC Projects from RFNC-VNIIEF devoted to improving laser beam quality. *Springer Proceedings in Physics*, 102, 291-301.

[132] Sacks, Z. S., Rozas, D., & Swartzlander, G. A. Jr. (1998). Holographic formation of optical-vortex filaments. *J. Opt. Soc. Am. B*, 15(8), 2226-2234.

[133] Kim, G. H., Jeon, J. H., Ko, K. H., Moon, H. J., Lee, J. H., & Chang, J. S. (1997). Optical vortices produced with a nonspiral phase plate. *Applied Optics*, 36(33), 8614-8621.

[134] Shvedov, V. G., Izdebskaya, Ya. V., Alekseev, A. N., & Volyar, A. V. (2002). The formation of optical vortices in the course of light diffraction on a dielectric wedge. *Technical Physics Letters*, 28(3), 256-260.

[135] Oemrawsingh, S. S. R., van Houwelingen, J. A. W., Eliel, E. R., Woerdman, J. P., Verstegen, E. J. K., Kloosterboer, J. G., & Hooft, G. W. (2004). Production and characterization of spiral phase plates for optical wavelengths. *Applied Optics*, 43(3), 688-694.

[136] Ganic, D., Gan, X., & Gu, M. (2002). Generation of doughnut laser beams by use of a liquid-crystal cell with a conversion efficiency near 100%. *Optics Letters*, 27(15), 1351-1353.

[137] Curtis, J. E., & Grier, D. G. (2003). Structure of Optical Vortices. *Phys. Rev. Lett*, 90(13), 133901.

[138] Fishman, A. I. (1999). Phase optical elements - kinoforms. *Soros Educ. J* [12], 76-83.

[139] Kamimura, T., Akamatsu, S., Horibe, H., et al. (2004). Enhancement of surface-damage resistance by removing subsurface damage in fused silica and its dependence on wavelength. *Jap. J. Appl. Phys*, 43(9), L1229-L1231.

[140] Sung, J. W., Hockel, H., Brown, J. D., & Johnson, E. G. (2006). Development of two-dimensional phase grating mask for fabrication of an analog-resist profile. *Applied Optics*, 45(1), 33-43.

[141] Swartzlander, G. A. Jr. (2006). Achromatic optical vortex lens. *Optics Letters*, 31(13), 2042-2044.

[142] Atuchin, V. V., Permyakov, S. L., Soldatenkov, I. S., & Starikov, F. A. (2006). Kinoform generator of vortex laser beams. *Proc. SPIE*, 6054, 1-4.

[143] Malakhov, Yu. I. (2006). ISTC projects devoted to improving laser beam quality. *Proc. SPIE*, 6346, 1-8.

[144] Malakhov, Yu. I., Atuchin, V. V., Kudryashov, A. V., & Starikov, F. A. (2009). Optical components of adaptive systems for improving laser beam quality. *Proc. SPIE*, 7131, 1-5.

[145] Southwell, W. H. (1980). Wave-front estimation from wave-front slope measurements. *J. Opt. Soc. Am*, 70(8), 998-1006.

[146] Vorontsov, M. A., Koryabin, A. V., & Shmalgausen, V. I. (1988). Controlled optical systems. *Moscow: Nauka*.

[147] Ghiglia, D. C., & Pritt, M. D. (1998). Two-dimensional phase unwrapping: theory, algorithms, and software. *New-York: Wiley.*

[148] Zou, W., & Zhang, Z. (2000). Generalized wave-front reconstruction algorithm applied in a Shack-Hartmann test. *Applied Optics,* 39(2), 250-268.

[149] Aksenov, V., Banakh, V., & Tikhomirova, O. (1998). Potential and vortex features of optical speckle fields and visualization of wave-front singularities. *Applied Optics,* 37(21), 4536-4540.

[150] Roggemann, M. C., & Koivunen, A. C. (2000). Branch-point reconstruction in laser beam projection through turbulence with finite-degree-of-freedom phase-only wave-front correction. *J. Opt. Soc. Am. A,* 17(1), 53-62.

[151] Aksenov, V. P., & Tikhomirova, O. V. (2002). Theory of singular-phase reconstruction for an optical speckle field in the turbulent atmosphere. *J. Opt. Soc. Am. A,* 19(2), 345-355.

[152] Tyler, G. A. (2000). Reconstruction and assessment of the least-squares and slope discrepancy components of the phase. *J. Opt. Soc. Am. A,* 17(10), 1828-1839.

[153] Le Bigot, E.-O., & Wild, W. J. (1999). Theory of branch-point detection and its implementation. *J. Opt. Soc. Am. A,* 16(7), 1724-1729.

[154] Aksenov, V. P., Izmailov, I. V., Kanev, F. Yu., & Starikov., F. A. (2005). Localization of optical vortices and reconstruction of wavefront with screw dislocations. *Proc. SPIE,* 5894, 1-11.

[155] Aksenov, V. P., Izmailov, I. V., & Kanev, F. Yu. (2005). Algorithms of a singular wavefront reconstruction. *Proc. SPIE,* 6018, 1-11.

[156] Aksenov, V. P., Izmailov, I. V., Kanev, F. Yu., & Starikov, F. A. (2006). Screening of singular points of vector field of phase gradient, localization of optical vortices and reconstruction of wavefront with screw dislocations. *Proc. SPIE,* 6162, 1-12.

[157] Aksenov, V. P., Izmailov, I. V., Kanev, F. Yu., & Starikov, F. A. (2006). Performance of a wavefront sensor in the presence of singular point. *Proc. SPIE,* 634133, 1-6.

[158] Aksenov, V. P., Izmailov, I. V., Kanev, F. Yu., & Starikov, F. A. (2007). Singular wave-front reconstruction with the tilts measured by Shack-Hartmann sensor. *Proc. SPIE,* 63463, 1-8.

[159] Aksenov, V. P., Izmailov, I. V., Kanev, F. Yu., & Starikov., F. A. (2008). Algorithms for the reconstruction of the singular wavefront of laser radiation: analysis and improvement of accuracy. *Quantum Electronics,* 38, 673-677.

[160] Starikov, F. A., Atuchin, V. V., Kochemasov, G. G., et al. (2005). Wave front registration of an optical vortex generated with the help of spiral phase plates. *Proc. SPIE,* 589, 1-11.

[161] Starikov, F. A., Aksenov, V. P., Izmailov, I. V., et al. (2007). Wave front sensing of an optical vortex. *Proc. SPIE*, 634, 1-8.

[162] Atuchin, V. V., Soldatenkov, I. S., Kirpichnikov, A. V., et al. (2004). Multilevel kinoform microlens arrays in fused silica for high-power laser optics. *Proc. SPIE*, 5481, 43-46.

[163] Leach, J., Keen, S., Padgett, M., Saunter, C., & Love, G. D. (2006). Direct measurement of the skew angle of the Poynting vector in helically phased beam. *Optics Express*, 14(25), 11919-11923.

[164] Starikov, F. A., Kochemasov, G. G., Kulikov, S. M., et al. (2007). Wave front reconstruction of an optical vortex by Hartmann-Shack sensor. *Optics Letters*, 32(16), 2291-2293.

[165] Starikov, F. A., Aksenov, V. P., Atuchin, V. V., et al. (2009). Correction of vortex laser beams in a closed-loop adaptive system with bimorph mirror. *Proc. SPIE*, 7131, 1-8.

[166] Starikov, F. A., Aksenov, V. P., Atuchin, V. V., et al. (2007). Wave front sensing of an optical vortex and its correction in the close-loop adaptive system with bimorph mirror. *Proc. SPIE*, 6747, 1-8.

[167] Bokalo, S. Yu., Garanin, S. G., Grigorovich, S. V., et al. (2007). Deformable mirror based on piezoelectric actuators for the adaptive system of the Iskra-6 facility. *Quantum Electronics*, 37(8), 691-696.

[168] Garanin, S. G., Manachinsky, A. N., Starikov, F. A., & Khokhlov, S. V. (2012). Phase correction of laser radiation with the use of adaptive optical systems at the Russian Federal Nuclear Center- Institute of Experimental Physics. *Optoelectronics, Instrumentation and Data Processing*, 48(2), 134-141.

[169] Starikov, F. A., Kochemasov, G. G., Koltygin, M. O., et al. (2009). Correction of vortex laser beam in a closed-loop adaptive system with bimorph mirror. *Optics Letters*, 34(15), 2264-2266.

[170] Soskin, M. S., Gorshkov, V. N., Vasnetsov, M. V., Malos, J. T., & Heckenberg, N. R. (1997). Topological charge and angular momentum of light beams carrying optical vortices. *Phys. Rev. A*, 56(5), 4064-4075.

Chapter 8

A Unified Approach to Analysing the Anisoplanatism of Adaptive Optical Systems

Jingyuan Chen and Xiang Chang

Additional information is available at the end of the chapter

1. Introduction

To improve the quality of a laser beam propagating in atmospheric turbulence or to improve the resolution of turbulence-limited optical systems, adaptive optics (AO) (Hardy 1998; Tyson 2011) has been developed. In classical AO systems, the compensation is realized by real-time detection of the turbulence-induced perturbations from a source (beacon) using a wave-front sensing device and then removing them by adding a conjugated item on the same path using a wave-front compensating device.

However, the perturbations caused by the beacon and the target may not be the same, so when the perturbations measured by the beacon are used to compensate the perturbations caused by the target, the compensation performance is degraded. These effects are referred to as anisoplanatism (Sasiela 1992). Anisoplanatic effects are present if there is a spatial separation between the target and beacon (Fried 1982), a spatial separation between the wave-front sensing and compensating apertures (Whiteley, Welsh et al. 1998), when time delays in the system cause the beacon phase and the target phase to be only partially corrected due to atmospheric winds or motion of the system components (Fried 1990) or when the beacon and target have different properties such as distributed size (Fried 1995; Stroud 1996) or wavelength (Wallner 1977), and so on.

Conventionally, all kinds of anisoplanatic effects are studied individually, assuming that they are statistically uncorrelated, and the total effects are obtained by summing them all together when necessary (Gavel, Morris et al. 1994). This conventional approach has a rich history dating back to the earliest days of AO technology and has obtained many good results. But this approach is very limited, because for actual applications of AO systems, many kinds of anisoplanatic effects exist simultaneously and are dependent on each other (Tyler 1994). It is increasingly obvious that these methods are inadequate to treat the diverse nature of new AO applications and the concept of anisoplanatism, and the associated analysis methods must be expanded to treat these new systems so their performance may be properly assessed.

Although anisoplanatism takes many forms, it can be quantified universally by the correlative properties of the turbulence-induced phase. Therefore, instead of investigating a particular form of anisoplanatism, this paper concentrates on constructing a unified approach to analyse general anisoplanatic effects and their effects on the performance of AO systems. For the sake of brevity, we will consider only the case of classic single-conjugate AO systems and not consider the case of a multi-conjugate AO system (Ragazzoni, Le Roux et al. 2005).

In section 2 the most general analysis geometry with two spatially-separated apertures and two spatially-separated sources is introduced. In section 3, we introduce the transverse spectral filtering method which will be used to develop the unified approach for anisoplanatism in this chapter and the general expression of the anisoplanatic wave-front variance will be introduced. In section 4, some special geometries will be analysed. Under these special geometries, the scaling laws and the related characteristic quantities widely used in the AO field, such as Fried's parameter, the Greenwood frequency, the Tyler frequency, the isoplanatic angle, the isokinetic angle, etc., can be reproduced and generalized. In section 5, two specific AO systems will be studied to illustrate the application of the unified approach described in this chapter. One of these systems is an adaptive-optical bi-static lunar laser ranging system and the other is an LGS AO system where, besides the tip-tilt components, the defocus is also corrected by the NGS subsystem. Simple conclusions are drawn in section 6.

2. General analysis geometry

In the development that follows, we will employ the geometry shown in Figure 1, which is introduced by Whiteley et al. (Whiteley, Roggemann et al. 1998). This geometry shows two apertures, including sensing aperture and compensation aperture, whose position vectors are given by $\vec{r}_s$ and $\vec{r}_c$. Two optical sources, including target and beacon, are located by position vectors $\vec{r}_t$ and $\vec{r}_b$, respectively. The position vectors of the two apertures and the two sources share a fixed coordinate system. A vertical atmospheric turbulence layer, located at altitude z, is also shown in Figure 1.

The projected separation of the aperture centres in this turbulence layer is given by

$$\vec{s}_z = \gamma_z \vec{d} + (\gamma_z - \alpha_z)\vec{r}_s + A_{tcz}\vec{r}_t - A_{bsz}\vec{r}_b \tag{1}$$

where $\vec{d} = \vec{r}_c - \vec{r}_s$ is the distance of two apertures, A_{bsz} and A_{tcz} are the layer scaling factors given by $A_{bsz} = [z - (\vec{r}_s \bullet \hat{z})]/[(\vec{r}_b - \vec{r}_s) \bullet \hat{z}]$ and $A_{tcz} = [z - (\vec{r}_c \bullet \hat{z})]/[(\vec{r}_t - \vec{r}_c) \bullet \hat{z}]$, while α_z and γ_z are propagating factors of beacon and target, and defined by $\alpha_z = 1 - A_{bsz}$, and $\gamma_z = 1 - A_{tcz}$.

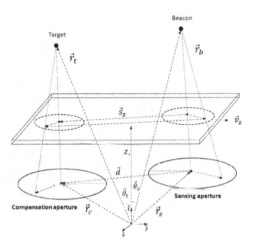

Figure 1. General geometry of the adaptive optical system

Under some hypotheses, these expressions can be further simplified. We suppose two aper-tures are at the same altitudes and select the centre of the sensing aperture as the origin of coordinates. We express the positions of target and beacon with the zenith angle and alti-tude as $(\vec{\theta}_t, L)$ and $(\vec{\theta}_b, H)$, respectively. We notice that in studying anisoplanatic effects, the offsets angular is very small in general (Welsh and Gardner 1991), i.e., $\vec{\theta} \ll 1$, then Eq. (1) is well approximated by

$$\vec{s}_z = \gamma_z \vec{d} + z\vec{\theta} \tag{2}$$

where $\vec{\theta} = \vec{\theta}_t - \vec{\theta}_b$ is the angular separation between target and beacon. At the same time, the propagating factors can be simplified to $\alpha_z = 1 - z/H$, and $\gamma_z = 1 - z/L$.

Further, if we consider delayed-time (τ) of the compensating process, then the projected separation can be expressed as

$$\vec{s}_z = \gamma_z \vec{d} + z\vec{\theta} + \vec{v}_z \tau \tag{3}$$

where $\vec{v}_z$ is the vector of wind velocity in this turbulent layer.

The above is the most general geometric relationship of AO systems. Depending on the conditions of application, more simple geometry can often be used to consider the aniso-planatism of AO systems. Some examples are showed in Figure 2. When the target is suf-ficiently bright, wave-front perturbation can be measured by directly observing the target. Thus an ideal compensation can be obtained and no anisoplanatism exists. This case is showed in Figure 2(a). In general, the target we are interested in is too dim to provide wave-front sensing, another bright beacon in the vicinity of the target must be used, as depicted in Figure 2(b). In this case, the so-called angular anisoplanatism exits (Fried 1982). In more general cases, a naturally existed object (NGS) cannot be find appropriately, to use AO systems, artificial beacons (LGS) must be created to obtained the wave-front perturbations (Happer, Macdonald et al. 1994; Foy, Migus et al. 1995). Then so called focal anisoplanatism (Buscher, Love et al. 2002; Muller, Michau et al. 2011) appears because of an altitude difference between LGS and target, as depicted in Figure 2(c). Figure 2(d) illustrates that a special anisoplanatism will be induced when a distributed source is used as the AO beacon because it is different from a pure point source (Stroud 1996). Distributed beacons are often occurred, for example, a LGS will wander and expand as a distributed source because of the effects of atmospheric turbulence when the laser is projected upward from the ground (Marc, de Chatellus et al. 2009). In Figure 2(e), the anisoplanatism induced by a separation of the wave-front sensing and compensation aperture is illustrated. With many applications, such as airborne lasers, the separated apertures are indispensable because of the moving platform (Whiteley, Roggemann et al. 1998). Figure 2(f) illustrates a hybrid case, in which many anisoplanatic effects coexist at the same time.

All these special anisoplanatic effects are degenerated cases and can be analysed under general geometry. In the following section, we will construct the general formularies of anisoplanatic variance under the most general geometry.

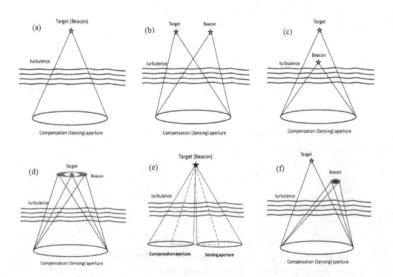

Figure 2. Some special cases of geometry and anisoplanatism. (a) ideal compensation, where the target is also used as the beacon; (b) angular anisoplanatism; (c) focal anisoplanatism; (d) extended beacon; (e) separated apertures; (f) hybrid beacon - many anisoplanatic effects existing at the same time.

3. Transverse spectral filtering method and general expressions of corrected (anisoplanatic) wave-front variance

Sasiela and Shelton developed a very effective analytical method to solve the problem of wave propagating in atmospheric turbulence (Sasiela 2007). This method uses Rytov's weak fluctuation theory and the filtering concept in the spatial-frequency domain for coordinates transverse to the propagation direction. In the most general case, the variance of a turbulence-induced phase-related quantity for the propagating waves, when diffraction is ignored, can be written as:

$$\sigma^2 = 2\pi k_0^2 \int_0^L dz (\int \Phi(\vec{\kappa}, z) f(\vec{\kappa}, z) d\vec{\kappa}) \tag{4}$$

where L is the propagation distance and k_0 is the space wave number, which when related to wavelength λ by $k_0 = 2\pi / \lambda$; $\Phi(\vec{\kappa}, z)$ is two-dimensional transverse power spectrum of fluctuated refractive-index at the plane vertical to the direction of wave propagation and $\vec{\kappa} = (\kappa, \varphi)$; $f(\vec{\kappa}, z)$ is the transverse spectral filter function related to this calculated quantity, whose explicit form can be determined by the corresponding physical processes.

For the atmospheric turbulence, the two-dimensional transverse power spectrum of fluctuated refractive-index can generally be written as:

$$\Phi(\vec{\kappa}, z) = 0.033 C_n^2(z) g(\kappa) \kappa^{-11/3} \tag{5}$$

where $C_n^2(z)$ is refractive-index structure parameter which is allowed to vary along the propagation path, and $g(\kappa)$ is the normalized spectrum. If $g(\kappa) = 1$, then the classic Kolmogrov spectrum is obtained.

Now substitute Eq. (5) into Eq. (4), and sequentially perform the integration of wave vector $\vec{\kappa} = (\kappa, \varphi)$ at the angular and radial components (Sasiela and Shelton 1993), then the variance reduces to

$$\sigma^2 = 0.4147 \pi k_0^2 \int_0^L dz C_n^2(z) I_F(z) \tag{6}$$

in which radial and angular integration can be written respectively as:

$$I_F(z) = \int_0^\infty F(\kappa, z) g(\kappa) \kappa^{-8/3} d\kappa \tag{7}$$

$$F(\kappa, z) = \frac{1}{2\pi} \int_0^{2\pi} f(\vec{\kappa}, z) d\varphi \tag{8}$$

To evaluate the integral Eq. (6), the expression of the filter function must be given. We will introduce the anisoplanatic filter function for general geometry illustrated in Figure 1. The anisoplanatic filter function can be created from some complex filter functions, describing the process related to the observed target and beacon respectively, by taking the absolute value squared of their difference.

Clearly, when $z \geq H$, the anisoplanatic filter function is

$$f(\vec{\kappa}, z) = |G(\gamma_z \vec{\kappa})|^2 \tag{9}$$

While when $z < H$, this item can be expressed as:

$$f(\vec{\kappa}, z) = |G(\gamma_z \vec{\kappa}) exp(i \vec{\kappa} \bullet \vec{s}_z) - G(\alpha_z \vec{\kappa}) G_s(\vec{\kappa}, z)|^2 \tag{10}$$

In above two equations, $G(\vec{\kappa})$ is the complex filter function corresponding to the wanted quantity, while $G_s(\vec{\kappa}, z)$ is a complex function which can describe the characteristic of the beacon (such as distributed or point-like). When writing this equation, we have supposed that the main physical processes are linear and their complex filter function can be cascaded to form the total filter functions.

Below we list some explicit expressions of complex filter functions.

The transverse complex filter function for a uniform, circular source with angular diameter θ_r, can be expressed as:

$$G_s(\vec{\kappa}, z) = 2J_1(\kappa\theta_r z)/(\kappa\theta_r z) \tag{11}$$

Here $J_n(\bullet)$ is the nth-order of Bessel function of the first kind; Similarly, the filter function for a Gaussian intensity distribution with $1/e$ radius θ_r, has a complex filter function

$$G_s(\vec{\kappa}, z) = exp[-(\kappa\theta_r z)^2/2] \tag{12}$$

We also notice that for a point-like beacon, the filter function is simply 1.

For the global phase, the complex filter function is

$$G_\phi(\vec{\kappa}, \vec{\rho}) = exp(i \vec{\kappa} \bullet \vec{\rho}) \tag{13}$$

The expression of the complex filter function for Zernike mode $Z(m,n)$ depends on its radial (n) and azimuthal (m) order. For $m = 0$, it can be written as:

$$G_{n,0}(\vec{\kappa}) = (-1)^{n/2} N_n(\vec{\kappa}) \tag{14}$$

For $m \neq 0$, it is given by

$$\left.\begin{array}{c} G_{n,m}^{x}(\vec{\kappa}) \\ G_{n,m}^{y}\vec{\kappa} \end{array}\right\} = i^{m}\sqrt{2}(-1)^{(n-m)/2}N_{n}(\vec{\kappa})\begin{cases} cos(m\varphi) \\ sin(m\varphi) \end{cases} \tag{15}$$

In previous two equations, D is the diameter of aperture and

$$N_{n}(\vec{\kappa}) = 2\sqrt{n+1}J_{n+1}(\kappa D/2)/(\kappa D/2) \tag{16}$$

By the above complex filter functions, the expressions of anisoplanatic filter functions of global phase and its Zernike modes can be established explicitly.

For the total phase, When $z \geq H$, from Eq. (9) and Eq. (13), it is

$$F_{\phi}(\kappa, z) = 1 \tag{17}$$

While when $z < H$, from Eq. (10) and Eq. (13), the result is

$$F_{\phi}(\kappa, z) = 1 - 2N_{0}((\gamma_{z} - \alpha_{z})\kappa)G_{s}(\kappa, z)J_{0}(s_{z}\kappa) + G_{s}^{2}(\kappa, z) \tag{18}$$

Similarly, the anisoplanatic filter functions for Zernike modes can also be established. For the case $z < H$, when $m = 0$, it can be given by the expression

$$F_{n,0}(\kappa, z) = N_{n}^{2}(\gamma_{z}\kappa) + N_{n}^{2}(\alpha_{z}\kappa)G_{s}^{2}(\kappa, z) - 2N_{n}(\gamma_{z}\kappa)N_{n}(\alpha_{z}\kappa)G_{s}(\kappa, z)J_{0}(s_{z}\kappa) \tag{19}$$

When $m \neq 0$, for the x, y component of Zernike mode, we can write their anisoplanatic filter functions as follows:

$$F_{n,m}^{x}(\kappa, z) = N_{n}^{2}(\gamma_{z}\kappa) + N_{n}^{2}(\alpha_{z}\kappa)G_{s}^{2}(\kappa, z) +$$
$$-2N_{n}(\gamma_{z}\kappa)N_{n}(\alpha_{z}\kappa)G_{s}(\kappa, z)\left[J_{0}(s_{z}\kappa) - (-1)^{m}J_{2m}(s_{z}\kappa)\right] \tag{20}$$

$$F_{n,m}^{x}(\kappa, z) = N_{n}^{2}(\gamma_{z}\kappa) + N_{n}^{2}(\alpha_{z}\kappa)G_{s}^{2}(\kappa, z)$$
$$-2N_{n}(\gamma_{z}\kappa)N_{n}(\alpha_{z}\kappa)G_{s}(\kappa, z)\left[J_{0}(s_{z}\kappa) + (-1)^{m}J_{2m}(s_{z}\kappa)\right] \tag{21}$$

It is easy to find that if we define a new quantity as follows:

$$F_{n,m}(\kappa, z) = F_{n,m}^{x}(\kappa, z) + F_{n,m}^{y}(\kappa, z) \tag{22}$$

then we can obtain

$$F_{n,m}(\kappa, z) = C_{m}\left[N_{n}^{2}(\gamma_{z}\kappa) + N_{n}^{2}(\alpha_{z}\kappa) G_{s}^{2}(\kappa, z) - 2 N_{n}(\gamma_{z}\kappa) N_{n}(\alpha_{z}\kappa) G_{s}(\kappa, z) J_{0}(s_{z}\kappa)\right] \tag{23}$$

Where C_m is a constant factor related to the azimuthal order m. If $m=0$, then $C_m=1$; otherwise $C_m=2$.

Similarly for the case $z \geq H$, the corresponding result is

$$F_{n,m}(\kappa, z) = C_m N_n^2(\gamma_z \kappa) \tag{24}$$

4. Some special cases

In the previous section the transverse anisoplanatic spectral filter functions for the general geometry of adaptive optical systems have been established. In this section we consider some special geometric cases, where asymptotic solutions of integrals can be obtained.

4.1. The anisoplanatism induced by separated apertures and its related characteristic distances

We first consider a simple case, where only the anisoplanatism induced by two separated apertures exists and the others are ignored. Let $\tau=0$, $\theta=0$, $G_s(\kappa, z)=1$, and $L = H = +\infty$ (i.e., $\gamma_z = \alpha_z = 1$), and taking into account the limitation $\lim_{x \to 0} N_0(x) = 1$, then Eq. (18) and Eq. (23) reduce to

$$F_\phi(\kappa, z) = 2[1 - J_0(\kappa d)] \tag{25}$$

$$F_{n,m}(\kappa, z) = 2C_m N_n^2(\kappa)[1 - J_0(\kappa d)] \tag{26}$$

The anisoplanatic phase variance is easily obtained. Substituting Eq. (25) into Eq. (6), and using the Kolmogrov spectrum, i.e., $g(\kappa)=1$, the integral is equal to

$$\sigma_\phi^2 = (d / d_0)^{5/3} \tag{27}$$

Here we have calibrated the variance with a new characteristic distance d_0, defined as

$$d_0 = 0.526 \, k_0^{-6/5} \mu_0^{-3/5} \tag{28}$$

This is about $1/3$ of the atmospheric coherence length $r_0 = (0.423 k_0^2 \mu_0)^{-3/5}$, Where μ_m represents the mth (full) turbulence moments. From Eq. (27), we find that the anisoplanatic variance induced by separated apertures meets the 5/3 power scaling law with the distance of separated apertures.

For AO systems, the piston phase variance is not meaningful and can be removed from the total variance. Their difference, i.e., the piston-removed phase variance, cannot be expressed analytically for arbitrary distances, while for very small and very large distance their asymptotic solutions can be found. We first calculate in these limitations the wave vector integral of the piston-removed anisoplanatic phase filter function $I_{\phi eff,aniso} = I_\phi - I_0$, which can be found easily from the Eq. (79) and (80) in Appendix with $n=0$.

When $d \gg D$, expanding $I_{\phi eff,aniso}$ to second order of (D/d), the result is

$$I_{\phi eff,aniso} \sim \left(\frac{D}{2}\right)^{5/3}\left[-\frac{4\,\Gamma(-5/6)\Gamma(7/3)}{\sqrt{\pi}\,\Gamma(17/6)\Gamma(23/6)} - \frac{\Gamma(1/6)}{4\,\Gamma(5/6)}\left(\frac{D}{d}\right)^{1/3}\right] \tag{29}$$

While $d \ll D$, expanding it to fourth-order of (d/D), the result is

$$I_{\phi eff,aniso} \sim -\frac{\Gamma(-5/6)}{\Gamma(11/6)}\left(\frac{D}{2}\right)^{5/3}\left(\frac{d}{D}\right)^{5/3} + \frac{\Gamma(7/3)\Gamma(-5/6)}{\sqrt{\pi}\,\Gamma(17/6)\Gamma(23/6)}\left(\frac{D}{2}\right)^{5/3}\left(\frac{d}{D}\right)^2\left[\frac{51425}{41472}\left(\frac{d}{D}\right)^2 - \frac{935}{144}\right] \tag{30}$$

On the other hand, the wave vector integral of the piston-removed phase filter function for a single wave beam is easy to find and can be expressed as:

$$I_{\phi eff,single} = -\frac{2\,\Gamma(-5/6)\Gamma(7/3)}{\sqrt{\pi}\,\Gamma(17/6)\Gamma(23/6)}\left(\frac{D}{2}\right)^{5/3} \tag{31}$$

From the above equations we find that in the limitation of $d \gg D$ the piston-removed anisoplanatic phase variance tends to be twice that of the piston-removed phase variance of a single wave. This is predictable, because when the separated distance of apertures is large enough, the correlation of waves from two separated aperture is gradually lost, and these beams are statistically independent of each other. We also find in the limitation of $d \ll D$ the piston-removed anisoplanatic phase variance remains the 5/3 power scaling law with the separated distance, which is same as that for the total phase in Eq. (27).

There are many ways to define a related characteristic distance. For an AO system, if the piston-removed anisoplanatic phase variance is greater than the same quantity for a single wave, that is to say

$$\sigma^2_{\phi eff,aniso}\big|_{d \ll D} > \sigma^2_{\phi eff,single} \tag{32}$$

Then the compensation is ineffective and the AO system is not needed. We can define the uncorrected distance d_{unc} of two separated apertures as the smallest distance satisfied above inequality. Using Eq. (30), Eq. (31) and Eq. (32), an approximation of this characteristic distance can be given by $d_{unc} = 0.828D$.

On the other hand, to achieve a better performance, the residual error of corrected wave must be small enough. Similar to the isoplanatic angle, we can define the isoplanatic distance as the separated distance of apertures at which the residual error is an exact unit. From the scaling law of Eq. (27), this distance is same as d_0, i.e., $d_{iso} = d_0$.

The above two characteristic distances (d_{unc} and d_{iso}) give different restrictions to an apertures-separated AO system. Other characteristic distances can also be defined. For example, for such an AO system, we can define the effective corrected distance (d_{eff}) as the separated distance of apertures at which the AO system can work effectively. Obviously this distance can be determined by the smaller of the above two characteristic distances, namely,

$$d_{eff} = Min\{d_{iso}, d_{unc}\} \tag{33}$$

In general, the inequality $d_{iso} < d_{unc}$ is always satisfied, so the effective corrected distance is $d_{eff} = d_{iso} = d_0$.

Similar to the above analysis and definitions for total phase, anisoplanatic variances and related characteristic distances can be determined for arbitrary Zernike modes. The final result is complex and can be expressed with generalized hypergeometric functions (Andrews 1998). In order to obtain a simpler close solution, we consider the limit case of very large or very small separating distance.

From Eq. (80), in the limitation $d \ll D$, the integral is approximately equal to

$$I_{n,m}(z) = C_m \frac{11}{2^{8/3}\sqrt{\pi}} \frac{\Gamma(7/3)\Gamma(n+1/6)(1+n)}{\Gamma(17/6)\Gamma(17/6+n)} \left(\frac{d}{D}\right)^2 D^{5/3} \tag{34}$$

Furthermore, performing the integration at the propagating path, the asymptotic value of the anisoplanatic phase variance for Zernike mode $Z(m,n)$ is obtained as follows:

$$\sigma_{n,m}^2 = 0.879 C_m k_0^2 \frac{(1+n)\Gamma(n+1/6)}{\Gamma(n+17/6)D^{1/3}} \mu_0 d^2 \tag{35}$$

If we defined the isoplanatic distance of the Zernike mode $Z(m,n)$ $d_{n,m;iso}$ as the distance satisfied the condition $\sigma_{n,m}^2 = 1$, then the variance can be calibrated as:

$$\sigma_{n,m}^2 = (d / d_{n,m;iso})^2 \tag{36}$$

This characteristic distance can be determined as follows:

$$d_{n,m;iso} = \frac{1.067}{k_0} \sqrt{\frac{\Gamma(n+17/6) D^{1/3}}{C_m \mu_0(n+1)\Gamma(n+1/6)}} \tag{37}$$

For a single beam, the expression corresponding to Eq. (34) is ($n \geq 1$)

$$I_{n,m;single}(z) = \frac{\Gamma(7/3)\Gamma(n-5/6)(1+n) C_m}{2^{2/3}\sqrt{\pi} \, \Gamma(17/6)\Gamma(n+23/6)} D^{5/3} \tag{38}$$

From Eq. (34) and Eq. (37), and another inequality similar to Eq. (32), the uncorrected distance of the Zernike mode $Z(m,n)$ can be defined as:

$$d_{n,m;unc} = 12D / \sqrt{11(6n-5)(6n+17)} \qquad (39)$$

Similarly, the effective distance of the Zernike mode $Z(m,n)$ can be defined as (at $n \geq 1$):

$$d_{n;eff} = \underset{m}{Min} \left(d_{n,m;iso}, d_{n,m;unc} \right) \qquad (40)$$

When the separated distance of the two apertures is smaller than this characteristic distance, the Zernike mode $Z(m,n)$ of turbulence-induced phase can be compensated effectively by the AO system. In Eq. (40), the Minimum operator is evaluated throughout all the field of m, so the result is no longer dependent on m.

In Figure 3, we show the typical values of the characteristic distances $d_{n;eff}$ defined above for the separated-apertures-induced anisoplanatism with $D=1.2m$. As a comparison with the total phase, the value of piston-removed quantity d_0 is also showed in the same figures at $n=0$.

In Figure 3(a), the relationship among $d_{n;eff}$ and the other two characteristic distances (for $d_{n,m;iso}$ and $d_{n,m;unc}$, their values also select the minimum in all the ms) are showed for $\lambda=532nm$. From this figure, we find that the isoplanatic distance is monotonous - increasing with the radial order of Zernike mode - while the uncorrected distance is decreasing with it. Therefore, the effective distance is determined by the isoplanatic distance when the radial order is small (such as for the tip-tilt, defocus, et al) and by the uncorrected distance when the radial order is large. We also find that the effective distances for small ns are usually greater than those for the (piston-removed) total phase, so when only a few low-level Zernike modes need to be compensated for, apertures with greater separated distance can be used.

Other sub-figures in Figure 3 show the effective distances for different compensational orders at different turbulent intensities and wavelengths. In Figure 3(b), four different turbulent intensities ($r_0=3cm$, $6cm$, $9cm$ and $12cm$ at reference wavelength of $\lambda=500nm$) are compared. In Figure 3(c), the effective distances for two different turbulence intensities ($r_0 = 5cm$ and $10cm$) and two different wavelengths ($\lambda=532nm$ and $1064nm$) are compared. We can find that the effective distances are smaller at stronger turbulences or smaller wavelengths.

In Figure 3(d), the relationships between the effective distances and turbulence intensities are showed for four different compensational orders ($n=1$, 2, 3, and 5) at $\lambda=532nm$. This shows that the effective (or uncorrected) distances are not related to the turbulence intensities for lager compensational orders, such as that for n=5.

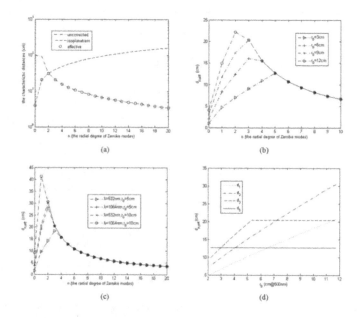

(a)   (b)

(c)   (d)

Figure 3. The characteristic distances for the anisoplanatism of separated apertures. (a) The relationship among three characteristic distances, $\lambda=532nm$; (b) The effective distances at for four different turbulent intensities, $\lambda=532nm$; (c) The effective compensational distances at different turbulent intensities and wavelengths; (d) The relationship between the effective distances and turbulence intensities for four different compensational orders, $\lambda=532nm$

4.2. The anglular anisoplanatism and related characteristic angles

Now we consider the geometry where only angular anisoplanatism exits. Let $d=0$, $\tau=0$, $\gamma_z=\alpha_z=1$, and $G_s(\kappa, z)=1$, then Eq. (18) and Eq. (23) reduce to

$$F_\phi(\kappa, z)=2[1 - J_0(\kappa\theta z)] \tag{41}$$

$$F_{n,m}(\kappa, z)=2C_m N_n^2(\kappa)[1 - J_0(\kappa\theta z)] \tag{42}$$

Substituting Eq. (41) into Eq. (6), and using the Kolmogrov spectrum, the result is $\sigma_\phi^2=(\theta / \theta_0)^{5/3}$, here θ_0 is the well-known isoplanatic angle defined as (Fried 1982) $\theta_0=(2.914k_0^2\mu_{5/3})^{-3/5}$.

Similarly, in the limitation of very small offset angle, i.e., $\theta z \ll D$, the effective corrected offset angle between beacon and target can be defined and determined by $\theta_{eff}=\theta_0$.

Using Eq. (42), the angular anisoplanatism of Zernike modes can also be calculated. The results can be expressed with the generalized hypergeometric functions, and in some limit conditions, a more compact expression can be obtained.

We consider the limitation of $\theta z \ll D$. Using Eq. (80) in Appendix, the angular anisoplanatism of Zernike mode $Z(m,n)$ can be expanded to the turbulence second-order structure constant moments and can be expressed as $(n \geq 1)$

$$\sigma_{n,m}^2 = (\theta / \theta_{n,m;iso})^2 \tag{43}$$

where the characteristic angle

$$\theta_{n,m;iso} = \frac{1.06665}{k_0 \mu_2^{1/2}} \sqrt{\frac{\Gamma(n + 17/6) \, D^{1/3}}{C_m(n + 1)\Gamma(n + 1/6)}} \tag{44}$$

can be defined as the isoplanatic angle for Zernike mode $Z(m,n)$, and it is the size of the offset-axis angle between the beacon and the target when the angular anisoplanatism of Zernike mode is unit rad^2.

When $n=1$ and $m=1$, the tip-tilt isoplanatic angle (also called isokinetic angle) is obtained. This characteristic angle can be expressed as:

$$\theta_{TA} = \theta_{1,1;iso} = 1.224\left(k_0^2 \mu_2 D^{-1/3}\right)^{-1/2} \tag{45}$$

This is consistent with the results in other research (Sasiela and Shelton 1993).

Similar to anisoplanatism of separated apertures, other characteristic angles can be defined and calculated. The uncorrected offset angle of $Z(m,n)$ can be expressed as:

$$\theta_{n,m;unc} = 12 \, D\sqrt{\mu_0 / \mu_2} / \sqrt{11(6n - 5)(6n + 17)} \tag{46}$$

and the effective offset angle of the n-order Zernike mode can be determined by

$$\theta_{n;eff} = \min_m \left(\theta_{n,m;iso}, \, \theta_{n,m;unc}\right) \tag{47}$$

In Figure 4, the typical values of the characteristic angles $\theta_{n;eff}$ defined above are showed at $D=1.2m$. In Figure 4(a), we compare the values for two different turbulent intensities ($r_0=5cm$, $10cm$) and two different wavelengths ($\lambda=532nm$, and $1064nm$). We can also find that the effective offset angles $\theta_{n;eff}$ for small ns are usually greater than those for the (piston-removed) total phase, the same as the characteristic quantities $d_{n;eff}$. In fact, this is one of main reasons that the use of LGS can partially solve the so-called "beacon difficulty", because a NGS may be find to correct the lower order modes of the turbulence-induced phase in a field far wider than that limited by the isoplanatic angle θ_0. Unlike $d_{n;eff}$, the effective offset angle $\theta_{n;eff}$ is not only dependent on aperture diameter D, but also turbulence intensity. Therefore, for higher-order Zernike modes, the effectively offset angle is also dependent on the turbulence intensity. In Figure 4(b), the relationships between effective offset angles and turbulence intensities are showed for four different compensational orders ($n=1, 2, 3,$ and 5) at $\lambda=532nm$.

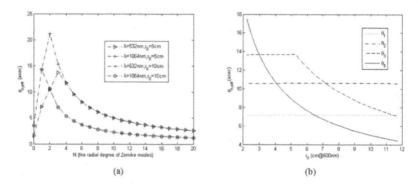

(a)   (b)

Figure 4. The characteristic angles of the angular anisoplanatism for separated beacon and target. (a) the effective offset angles at different turbulent intensities and wavelengths; (b) the relationship between effective offset angles and turbulence intensities for four different compensational orders at $\lambda=532nm$

4.3. The time-delayed anisoplanatism and related characteristic quantities

When $d=0$, $\theta=0$, $\gamma_z=\alpha_z=1$, and $G_s(\kappa, z)=1$, then there is only time-delayed anisoplanatism. Now Eq. (18) and Eq. (23) reduce to

$$F_\phi(\kappa, z)=2[1 - J_0(\kappa\, v_z\tau)] \tag{48}$$

$$F_{n,m}(\kappa, z)=2C_m N_n^2(\kappa)[1 - J_0(\kappa\, v_z\tau)] \tag{49}$$

Using Eq. (48) and $g(\kappa)=1$ to perform the integration in Eq. (6), the total phase anisoplanatism variance can be expressed as $\sigma_\phi^2=(\tau/\tau_0)^{5/3}$, where τ_0 is normally-defined atmospheric coherence time and equal to $\tau_0=(2.913k_0^2v_{5/3})^{-3/5}$, and v_n is the nwth velocity moments of atmospheric turbulence defined by $v_n=\int_0^L dzC_n^2(z)v^m(z)$. This characteristic quantity τ_0 is related to the Greenwood frequency. For a single-poles filter (controller), the variance of compensated phase can be scaled as $\sigma_\phi^2=(f_0/f_{3db})^{5/3}$, where f is the effective control bandwidth of AO system and f_0 is the Greenwood frequency, defined by $f_0=(0.103k_0^2v_{5/3})^{3/5}$. We can easily find there is a simple relationship between these two characteristic quantities:

$$f_0 \approx 0.134 / \tau_0 \tag{50}$$

as is first noted by Fried (Fried 1990).

Similarly, in the limitation $v_z \tau \ll D$, the effective corrected time can be defined and determined by $\tau_{eff} = \tau_0$. For arbitrary Zernike mode of phase, from Eq. (80), when we consider the second order approximation, the anisoplanatic variance is equal to

$$\sigma_{n,m}^2 = 0.879 \frac{(1+n)\Gamma(n+1/6)}{\Gamma(n+17/6)} C_m k_0^2 D^{-1/3} v_2 \tau^2 \tag{51}$$

Using the isoplanatic time $\tau_{n,m;iso}$ satisfied $\sigma_{n,m}^2 = 1$ to rescale, then the variance can be expressed as:

$$\sigma_{n,m}^2 = (\tau / \tau_{n,m;iso})^2 \tag{52}$$

and its expression is

$$\tau_{n,m;iso} = \frac{1.06665}{k_0 v_2^{1/2}} \sqrt{\frac{\Gamma(n+17/6) \, D^{1/3}}{C_m (n+1)\Gamma(n+1/6)}} \tag{53}$$

Similar to Greenwood frequency, we can apply Eq. (50) to define a characteristic frequency related to the isoplanatic time in Eq. (53) as follows:

$$f_{n,m;iso} = 0.1256 k_0 v_2^{1/2} \sqrt{\frac{C_m(n+1)\Gamma(n+1/6)}{\Gamma(n+17/6) \, D^{1/3}}} \tag{54}$$

This is the characteristic frequency using a single-poles filter to compensate for the Zernike mode Z(m,n) of the turbulence-induced phase.

Further, the effective correction time of arbitrary n-order Zernike model of phase can be defined as:

$$\tau_{n;eff} = \min_m \left(\tau_{n,m;iso}, \tau_{n,m;unc} \right) \tag{55}$$

where the uncorrected time can be expressed as

$$\tau_{n,m;unc} = 12 \, D\sqrt{\mu_0 / v_2} / \sqrt{11(6n-5)(6n+17)} \tag{56}$$

When using an AO system with a time delay exceeding this characteristic time to compensate for the n-order Zernike model of phase, the compensation is ineffective.

The characteristic quantities $\tau_{n;\text{eff}}$ are similar to $\theta_{n;\text{eff}}$ and $d_{n;\text{eff}}$. In Figure 5, we show some typical values of the characteristic angles $\tau_{n;\text{eff}}$.

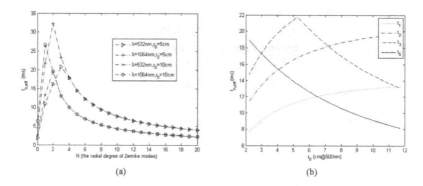

(a)　　　　　　　　　　　　　　　(b)

Figure 5. The characteristic times for the time-delay anisoplanatism. (a) The effective times at different turbulent intensities and wavelengths; (b) The relationship between effective times and turbulence intensities for different compensational orders at $\lambda=532nm$.

When $n = 1$ and $m = 1$, the isoplanatic times or the characteristic frequencys for the tip-tilt component of the turbulence-induced phase are obtained as: $\tau_{1,1;iso}=\left(0.668k_0^2 v_2 D^{-1/3}\right)^{-1/2}$ or $f_{1,1;iso}=0.4864\,\lambda^{-1}v_2^{1/2}\,D^{-1/6}$. It should be noted that these results are slightly different with others. In many studies, the tilt anisoplanatic variances are calibrated as: $\sigma_\alpha^2=\frac{1}{5}\left(\frac{\lambda}{D}\right)^2\left(\frac{\tau_{st}}{\tau_{0t}}\right)^2$ or $\sigma_\alpha^2=\left(\frac{\lambda}{D}\right)^2\left(\frac{f_T}{f_{3db}}\right)^2$, Where the characteristics time (Parenti and Sasiela 1994) and frequency (Tyler 1994) are defined by $\tau_{0t}=\left(0.512k_0^2 v_{-1/3}^{8/15} v_{14/3}^{7/15} D^{-1/3}\right)^{-1/2}$ or $f_T=0.368\,\lambda^{-1}v_2^{1/2}\,D^{-1/6}$. These results are slightly different from ours because different methods of series expanding are used. However, the differences are minor and our expressions have simpler forms and are more convenient to use.

4.4. The focal anisoplanatism

If the altitudes of beacon and target are different, then focal anisoplanatism appears. When other anisoplanatic effects are neglect (i.e., $\theta=0$, $d=0$, $\tau=0$, $L = +\infty$, $G_s(\kappa, z)=1$), the anisoplanatic filter function below the beacon are simplified to

$$F_\phi(\kappa, z)=2\left[1-\frac{H}{\kappa Dz}J_1\left(\frac{\kappa Dz}{2H}\right)\right]$$ (57)

$$F_{n,m}(\kappa, z)=C_m\left[N_n(\kappa)-N_n(\alpha_z\kappa)\right]^2$$ (58)

Substituting Eq. (57) into Eq. (6), the anisoplanatic variance for total phase is given by $\sigma_\phi^2 = 0.5 k_0^2 \mu_{5/3}^- (D/H)^{5/3}$, here μ_m^- is the mth lower turbulence moment, defined by $\mu_m^- = \int_0^H C_n^2(z) z^m dz$.

Similarly, using Eq. (58) the anisoplanatic variance of Zernike mode $Z(m,n)$ can also be calculated. In order to obtain a more simple close solution, we consider the limit case of a very high altitude beacon, i.e., $H \gg z$. From Eq. (81), when the second-order small quantities are retained, the anisoplanatic variance for $Z(m,n)$ can be approximated by

$$\sigma_{n,m}^2 = \frac{3.317}{6237} \frac{(1+n)\Gamma^2(-8/3)[108n(n+2)-55]}{\Gamma(-10/3)\Gamma(n+23/6)/\Gamma(n-5/6)} C_m k_0^2 \mu_2^- D^{5/3}/H^2 \tag{59}$$

By this expression, the first two components, i.e., the anisoplanatic variances of the piston and tip-tilt, can be obtained immediately as follows: $\sigma_P^2 = \sigma_{0,0}^2 = 0.0834 k_0^2 D^{5/3} \mu_2^- / H^2$ and $\sigma_T^2 = \sigma_{1,1}^2 = 0.3549 k_0^2 D^{5/3} \mu_2^- / H^2$.

When analyzing a LGS AO system with a telescope aperture of diameter D, it is useful to express the anisoplanatic variance by $\sigma_\phi^2 = (D/d_e)^{5/3}$, where the characteristic quantity d_e is a measure of effective diameter of the LGS AO system (Tyler 1994) (i.e., a telescope with a diameter equal to d_e will have 1 rad of rms wave-front error). Considering the fact that for a LGS system piston is meaningless and tip-tilt is non-detectable (Rigaut and Gendron 1992; Esposito, Ragazzoni et al. 2000), then an approximated value of d_e can be obtained by

$$d_e = \{k_0^2[0.5\mu_{5/3}^-/H^{5/3} - 0.4383\mu_2^-/H^2]\}^{-3/5} \tag{60}$$

We can further consider the effect of turbulence above the beacon. From Eq. (17) and Eq. (24), the filter functions for the total phase and its Zernike mode $Z(m,n)$ above the beacon are

$$F_\phi(\kappa, z) = 1 \tag{61}$$

$$F_{n,m}(\kappa, z) = C_m N_n^2(\kappa) \tag{62}$$

Therefore the anisoplanatic filter function of the partial phase in which the components of the piston and tip-tilt are removed can be expressed as:

$$F_{eff,up}(\kappa, z) = F_\phi(\kappa, z) - F_{0,0}(\kappa, z) - F_{1,1}(\kappa, z) \tag{63}$$

Performing the integration Eq. (6), the corresponding variance is obtained as

$$\sigma_{eff,up}^2 = 0.0569 k_0^2 \mu_0^+ D^{5/3} \tag{64}$$

Where μ_0^+ is the mth upper turbulence moment, defined by $\mu_m^+ = \int_H^\infty C_n^2(z)z^m dz$. So when consider the effect of turbulence above the beacon, the effective diameter can be expressed approximately as

$$d_e = \left\{ k_0^2 \left[0.0569\mu_0^+ + 0.5\mu_{5/3}^- / H^{5/3} - 0.4383\mu_2^- / H^2 \right] \right\}^{-3/5} \tag{65}$$

This is the same result as that obtained in other research (Sasiela 1994).

4.5. The anisoplanatism induce by an extended beacon

We now consider the anisoplanatic effect induced by a distributed beacon and neglect other anisoplanatic effects. Let $d=0$, $\theta=0$, $\tau=0$, and $\gamma_z = \alpha_z = 1$, then Eq. (18) and Eq. (23) are reduced to

$$F_\phi(\kappa, z) = [1 - G_s(\kappa, z)]^2 \tag{66}$$

$$F_{n,m}(\kappa, z) = C_m N_n^2(\kappa)[1 - G_s(\kappa, z)]^2 \tag{67}$$

Substituting above two equations into Eq. (6) and performing the integration, the anisoplanatic variance of the total phase and its Zernike components can be obtained. Below we give the corresponding results for a Gaussian distributed beacon and Kolmogrov's turbulent spectrum, i.e., using Eq. (12) and $g(\kappa)=1$.

For the total phase, the integration can easily be obtained. The result is $\sigma_\phi^2 = 0.5327\,\theta_r^{5/3}\mu_{5/3} = (0.3608\,\theta_r / \theta_0)^{5/3}$, here μ_m is the mth turbulence moment, and θ_0 is atmospheric isoplanatic angle. Obviously, the result is similar to the classic 5/3 power scaling law for angular anisoplanatism.

For Zernike component $Z(m,n)$, we consider the limit case of very big θ_r, i.e., $\theta_r z \gg D$. From Eq. (82), the approximate results expanding to the second order turbulence moment can be obtained. Here, we only list the first two components (i.e., the anisoplanatic variances of piston and tip-tilt) as follows: $\sigma_P^2 = \sigma_{0,0}^2 = 0.5327\theta_r^{5/3}\mu_{5/3} - 0.4369\,D^{5/3}\mu_0$ and $\sigma_T^2 = \sigma_{1,1}^2 = 0.3799\,D^{5/3}\mu_0$.

5. Two examples for hybrid anisoplanatism

To illustrate the application of the unified approach described in this chapter, we will study two special AO systems as examples in this section. In these examples many anisoplanatic effects exist at the same time, so no analytical solution for anisoplanatic variances can be obtained - only numeric results.

To calculate the anisoplanatic variances, we use the Hufnagel-Valley model:

$$C_n^2(z) = A \exp\left(-\frac{z}{10^2}\right) + \frac{2.7}{10^{16}} \exp\left(-\frac{z}{1500}\right) + \frac{5.94}{10^3}\left(\frac{w}{27}\right)^2\left(\frac{z}{10^5}\right)^{10} \exp\left(-\frac{z}{10^3}\right) \tag{68}$$

where w is the pseudo-wind, and the altitude z expressed in meters. The turbulence strength is usually changed by a variation of the w term or A, the parameter to describe the turbulence strength at the ground. At the same time, the modified von Karman spectrum

$$g(\kappa) = \left[1 + (\kappa_o/\kappa)^2\right]^{-11/6} \exp\left[-(\kappa/\kappa_i)^2\right] \tag{69}$$

will be use. Where κ_o and κ_i are the space wave numbers corresponding to the outer scale and the inner scale of the atmospheric turbulence, respectively. To consider the effect of time-delay, we use the Bufton wind model

$$v_z = v_g + 30 \exp\left[-(z - 9400)^2/4800^2\right] \tag{70}$$

Where v_g is the wind speed on the ground.

5.1. An adaptive-optical bi-static Lunar Laser Ranging (LLR) system

Although the technique of Lunar Laser Ranging (LLR) is one of most important methods to modern astronomy and Earth science, it is also a very difficult task to develop a successful LLR system (Dickey, Bender et al. 1994). One of the main reasons is that the quality of the outgoing laser beams deteriorates sharply due to the effect of atmospheric turbulence, including the wandering, expansion, and scintillation. To mitigate these effects of atmospheric turbulence and improve the quality of laser beams, one can use AO systems to compensate the outgoing beams (Wilson 1994; Riepl, Schluter et al. 1999). In this section we will study the anisoplanatism of a special adaptive optical bi-static LLR system in which the receiving aperture is also used to measure the turbulence-induced wave-front and the outgoing beam is compensated by the conjugated wave-front measured by this aperture. It is a concrete application of the unified approach described in this paper.

For this special AO system, two apertures and the useful point-like beacon (Aldrin, Collins, et al.) and the targets (Apollo 11, Apollo 15, et al.) are separated, so the anisoplanatism is hybrid. Let $G_s(\kappa, z)=1$, $L = H$ ($=3.8\times10^8 m$), and denote respectively the offset distance and angle of apertures and sources as d and θ, then the anisoplanatic filter function in altitude z are reduced to

$$F_\phi(\kappa, z) = 2[1 - J_0(s_z\kappa)] \tag{71}$$

$$F_{n,m}(\kappa, z) = 2C_m N_n^2(\gamma_z\kappa)[1 - J_0(s_z\kappa)] \tag{72}$$

where $\gamma_z = 1 - z/L$, $s_z = \gamma_z d + z\theta + v_z \tau$, and the corrected time delay has been considered. Using above Equations, the variances can be computed easily, but the results can be expressed by higher transcendental functions with no simpler expressions existing.

In Figure 6, we show the anisoplanatic variances when turbulence-induced wave-fronts are compensated to different Zernike orders.

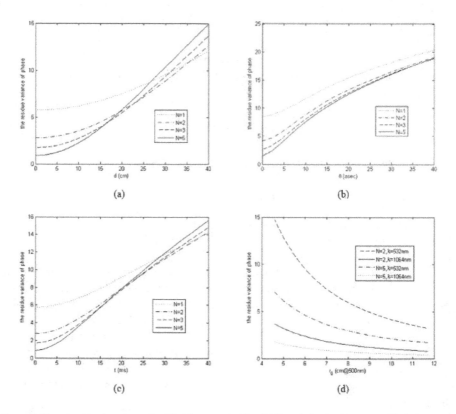

Figure 6. The anisoplanatic variances for LLR AO system, $D=1.2m$. (a) The relationship between residual phase variance and separated distances for four different compensational orders at $\lambda=532nm$ and $r_0=10cm$; (b) the relationship between residual phase variance and offset angles; (c) the relationship between residual phase variance and corrected time-delays for four different compensational orders at $\lambda=532nm$ and $r_0=10cm$; (d) the relationship between residual phase variance and turbulent intensities for two different compensational orders ($n=2$ and 5) and two different wavelengths ($\lambda=532nm$ and $1064nm$)

In the first three sub-graphs, the relationships between the anisoplanatic variances and some important parameters (separation distance of apertures, offset angle of sources, time-delay of the correcting process) are also showed respectively. From Figure 6(a), we can see the variances usually monotonously increase with the separated distance. We can also see that increasing the corrected order the variance will decrease when the separated distance is small, but it will not decrease when the separated distance is increased to a certain scale. This is because the effective distances $d_{n;eff}$ are smaller at larger orders, as has been showed in Figure 3. A similar conclusion can be drawn for the offset angle of sources from Figure 6(b) and for the time delay of the correcting process from Figure 6(c).

In Figure 6 (d), the relationship between anisoplanatic variance and turbulence intensity are showed for two wavelengths ($\lambda=532nm$ and $1064nm$) and two corrected orders ($n=2$ and 5). In this case, all three anisoplanatic effects (angular, time-delayed and that induced by separated apertures) exist at the same time and the corresponding parameters are selected as $d=5cm$, $\theta=2"$, $\tau=2ms$.

5.2. A special LGS AO system: Defocus corrected by the NGS subsystem

A laser beacon is insensitive to full-aperture tilt because the beam wanders on both the upward and the downward trips through the atmosphere, so currently when using LGS AO systems other NGS subsystems are usually used to sense and correct wave-front tilt. All other Zernike modes except tip-tilt can be corrected by LGS subsystems, but the corrected performance is limited by the focal anisoplanatism. Besides tip-tilt, the defocus (or focus) mode is another main component of the turbulence-induced phase and decreasing the focal anisoplanatism of the defocus component is very important (Esposito, Riccardi et al. 1996; Neyman 1996). In this section, we consider the performance of a special kind of LGS AO system, in which, besides the overall tilt, the focus mode can also be sensed and corrected by the NGS subsystems. Using this special LGS AO system, the focal anisoplanatism of the defocus mode can be reduced further.

We concentrate on the relationship between the focal and angular anisoplanatism of the defocus mode, and neglect the effects induced by time-delay and separated aperture. We also neglect the correlation between LGS and NGS subsystem, and suppose them to be statistically independent of each other. Then the anisoplanatic filter functions for the NGS subsystem are reduced to

$$F_{\phi,N}(\kappa, z)=1-2G_{s,N}(\kappa, z)J_0(\kappa z\theta_N)+G_{s,N}^2(\kappa, z) \tag{73}$$

$$F_{n,m;N}(\kappa, z) = C_m N_n^2(\gamma_z\kappa)\left[1-2G_{s,N}(\kappa, z)J_0(\kappa z\theta_N)+G_{s,N}^2(\kappa, z)\right] \tag{74}$$

While for the LGS subsystem, under the LGS beacon, the results reduce to

$$F_{\phi,L}(\kappa, z)=1-2N_0((\gamma_z-\alpha_z)\kappa)G_{s,L}(\kappa, z)J_0(\kappa z\theta_L)+G_{s,L}^2(\kappa, z) \tag{75}$$

$$F_{n,m;L}(\kappa, z)=C_m\left[N_n^2(\gamma_z\kappa)+N_n^2(\alpha_z\kappa)G_{s,L}^2(\kappa, z)-2N_n(\gamma_z\kappa)N_n(\alpha_z\kappa)G_{s,L}^2(\kappa, z)J_0(\kappa z\theta_L)\right] \tag{76}$$

Those above the LGS beacon are same as Eq. (17) and Eq. (24).

In above equations, γ_z, θ_N, and $G_{s,N}$ (or α_z, θ_L, $G_{s,L}$) are main related parameters of anisoplanatic effect, and they are the propagating factor, the offset angle, and the filter function of the NGS (or LGS), respectively. Here we have supposed that the altitude of NGS is same as that of the target.

Using these filter functions, the effective anisoplantic variance for this particular LGS AO system can be calculated and expressed as follows:

$$\sigma_{eff}^2 = \left(\sigma_{1,1;N}^2 + \sigma_{2,0;N}^2\right) + \left[\sigma_{\phi,L}^2 - \left(\sigma_{0,0;L}^2 + \sigma_{1,1;L}^2 + \sigma_{2,0;L}^2\right)\right] \tag{77}$$

In this equation, the first two items in parentheses are the contribution of the NGS subsystem, describing the anisoplanatism of tip-tilt and defocus modes respectively. While the items in brackets are the contribution of the LGS subsystem, and the four items are the variance of the total phase, the piston, the tip-tilt and the defocus mode, sequentially. As a comparison, the effective anisoplanatic variance for a usual LGS AO system, in which only tip-tilt mode can be sensed and corrected by the NGS subsystem, can be expressed as:

$$\sigma_{eff}^2 = \sigma_{1,1;N}^2 + \left[\sigma_{\phi,L}^2 - \left(\sigma_{0,0;L}^2 + \sigma_{1,1;L}^2\right)\right] \tag{78}$$

Obviously, for this special LGS AO system, the contribution of the defocus mode to the effective anisoplanatic variance comes from the NGS system, i.e., $\sigma_{2,0;N}^2$, while for a usual LGS AO system, it comes from the LGS subsystem.

Below we give some numerical results. We mainly study the changes of the anisoplanatic variance with some control parameters, including the altitudes (L and H), the offset angles (θ_N and θ_L), and the angular width (for Gaussian sources: $\theta_{r,N}$ and $\theta_{r,L}$) of the NGS and LGS sources. Some typical results are showed in the figures below. In our calculation, the altitude of the target L is selected as $500km$ (a typical value for a LEO satellite), and the wavelength as $1.315\mu m$.

In Figure 7(a) and Figure 7(b), the changes of the anisoplanatic variance with the angular widths and the offset angles of the beacons are given. In this case the invalid piston component of variance has been removed. In these figures, we also compare the values for three different altitudes of beacons, including a NGS ($H=L=500km$) and two kinds of LGSs with altitude $H=15km$ and $H=90km$ respectively. It is easy to see that the variances generally increase with the offset angles and the angular widths of the beacons. But there is some minor difference for the beacon size: the variance first decrease as beacon size increases, then it increases. We can also see that the changes are more obvious when the altitudes of the beacons are larger, for example, we can see the variance changes from 0.1 to 1.6 rad^2 when the offset angles changes from 0 to $10''$ for NGS, but there are nearly no changes for $15km$ Rayleigh LGS, as showed in Figure 7(b).

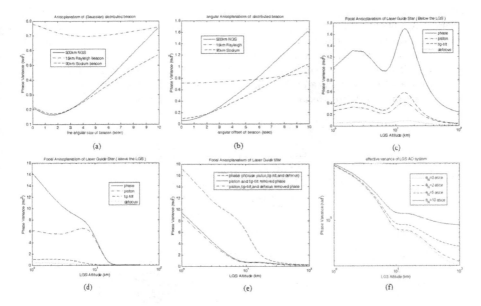

Figure 7. (a) Anisoplanatism of distributed beacon; (b) angular anisoplanatism; (c) the focal anisoplanatism (below the beacon); (d) focal anisoplanatism (above the beacon); (e) focal anisoplanatism (sum); (f) effective variance.

In Figure 7(c) and (d), the components of anisoplanatic variance below and above the beacon, are given respectively. The values for the total phase and its first three components (piston, tilt and defocus) are showed altogether. In Figure 7(e), the variances for the total phase, the piston and tip-tilt removed phase, and the piston and tip-tilt and defocus removed phase, are showed respectively. When the altitudes of the beacon are more than *20 km* the variances are almost the same as the results of the NGS. In Figure 7(f), the effective anisoplanatic variances expressed by Eq. (77) are showed for three different offset angles of NGS.

For the special LGS AO system, the anisoplanatic variance of defocus comes from NGS subsystem and not from the LGS subsystems as usual LGS AO systems. In Figure 8, we compare the values of these two variances and the relationship between the altitude of LGS and the offset angle of NGS. The transverse coordinates are magnitudes of the variances. The solid line describes the change of the defocus variances with the altitude of LGS and the altitude of LGS is showed in the left longitudinal coordinates. Similarly, the dotted line describes the change of the defocus variances with the offset angle of NGS and the offset angle of NGS is showed in the right longitudinal coordinates.

From this figure the value of the LGS altitude and the NGS offset angle, having the same value of the variance, can be read directly and some operational conclusions can be drawn.For example, for a Rayleigh LGS (with an altitude of *10km* to *20km*) the anisoplanatic variance of the focus component has the value between *0.08* to *0.1 rad²*, same as that for a NGS with the offset angle between 8" and 9". Similarly, the sodium LGS (with altitude of

90km) correspond to the offset angle of NGS between 2" and 3". It is also easy to see that the variance is a monotonically increasing function of the NGS offset angle and a almost monotonically decreasing function of the LGS altitude. Therefore, if the NGS offset angle is smaller or 0" (such as directly imaging of a bright satellite) using NGS to correct the defocus component, the variance is smaller. Otherwise, when the NGS offset angle is larger (for example, when projecting laser beams to a LEO satellite, the advance angle about 10" must be considered) using sodium LGS to correct the defocus the variance is smaller.

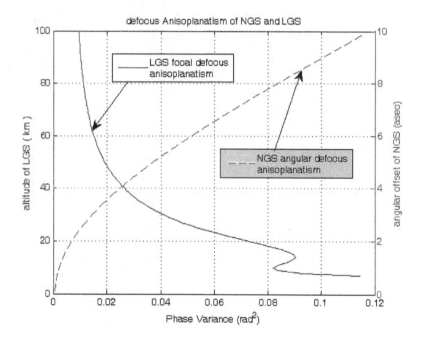

Figure 8. The anisoplanatism of the defocus component

6. Summary

Using transverse spectral filtering techniques we reconsider the anisoplanatism of general AO systems. A general but simple formula was given to find the anisoplanatic variance of the turbulence-induced phase and its arbitrary Zernike components under the general geometry of AO systems. This general geometry can describe most kinds of anisoplanatism appearing in currently running AO systems, including angular anisoplanatism, focal anisoplanatism and that induced by distributed sources or separated apertures, and so on. Under some special geometry, close-form solutions can be obtained and are consistent with classic results, which prove the effectiveness and universality of the general formula constructed in this chapter. We also give some numerical results of hybrid anisoplanatism under some more complex geometry.

Appendix

Here we give some expressions describing the integrations of the anisoplanatic filter function $F_{n,m}(\kappa, z)$ for the Zernike mode $Z(m,n)$ with respect to the radial component of the wave vector, i.e.,

$$I_n(z) = \int_0^\infty d\kappa \, F_{n,m}(\kappa, z)\kappa^{-8/3}$$

Expressions used in section 4.1.- 4.3.

When the filter function is equal to

$$F_{n,m}(\kappa, z) = 2N_n^2(\gamma_z\kappa)[1 - J_0(s_z\kappa)]$$

The results are determined by the sizes of s_z and $\gamma_z D$. If $|s_z| \geq \gamma_z D$, then

$$I_n(z) = (n+1)\Gamma\left(n - \frac{5}{6}\right)(\gamma_z D)^{5/3}\left[\frac{2^{1/3}\Gamma(7/3)/\Gamma(17/6)}{\sqrt{\pi}\,\Gamma(n+23/6)} - \right.$$

$$\left. -_3 \frac{F_2(n-5/6, n-5/6, n+3/2; n+2, 2n+3; |\gamma_z D/s_z|^2)}{2^{2n+5/3}\Gamma(-n+11/6)\Gamma^2(n+2)|\gamma_z D/s_z|^{-2n+5/3}}\right]$$

Further, if $|s_z| \gg \gamma D$, an asymptotic series of small parameter $|\gamma_z D/s_z|$ can be found. When expanding to second-order, the results are

$$I_n(z) = (1+n)\Gamma\left(n - \frac{5}{6}\right)(2D\gamma_z)^{\frac{5}{3}}\left\{\frac{\Gamma\left(\frac{7}{3}\right)/\Gamma\left(n+\frac{23}{6}\right)}{2^{4/3}\sqrt{\pi}\,\Gamma\left(\frac{17}{6}\right)} - -\frac{2\,|\gamma_z D/s_z|^{2n+1/3}}{\Gamma\left(\frac{11}{6} - n\right)[2^n\Gamma(2+n)]^2}\left[\frac{\left(n - \frac{5}{6}\right)^2}{n+2} + \frac{2^{-13/3}}{(\gamma_z D/s_z)^2}\right]\right\} \quad (79)$$

if $|s_z| < \gamma_z D$, then

$$I_n(z) = 2^{1/3}(n+1)(\gamma_z D)^{5/3}\left\{\frac{\Gamma\left(\frac{7}{3}\right)\Gamma\left(n - \frac{5}{6}\right)}{\sqrt{\pi}\,\Gamma\left(\frac{17}{6}\right)\Gamma\left(n + \frac{23}{6}\right)} - \right.$$

$$-\frac{\Gamma\left(\frac{7}{3}\right)\Gamma\left(n - \frac{5}{6}\right)}{\sqrt{\pi}\,\Gamma\left(\frac{17}{6}\right)\Gamma\left(n + \frac{23}{6}\right)}\,_3F_2\left(-\frac{11}{6}, -n - \frac{17}{6}, n - \frac{5}{6}; -\frac{4}{3}, 1; \left|\frac{s_z}{\gamma_z D}\right|^2\right) -$$

$$\left. -\frac{\Gamma\left(-\frac{7}{3}\right)}{\pi\,\Gamma\left(\frac{10}{3}\right)}\left|\frac{s_z}{\gamma_z D}\right|^{14/3}\,_3F_2\left(\frac{1}{2}, -n - \frac{1}{2}, n + \frac{3}{2}; \frac{10}{3}, \frac{10}{3}; \left|\frac{s_z}{\gamma_z D}\right|^2\right)\right\}$$

Further if $|s_z| \ll \gamma_z D$, the second-order asymptotic expansions of parameter $\left|\frac{s_z}{\gamma_z D}\right|$ are

$$I_n(z) = \frac{(1+n)\Gamma\left(\frac{7}{3}\right)}{\sqrt{\pi}\ \Gamma\left(\frac{17}{6}\right)}\left(\frac{\gamma_z D}{2}\right)^{\frac{5}{3}}\left|\frac{s_z}{\gamma_z D}\right|^2\left[\frac{11\ \Gamma\left(n+\frac{1}{6}\right)}{2\ \Gamma\left(n+\frac{17}{6}\right)} - \frac{55\ \Gamma\left(n+\frac{7}{6}\right)}{16\ \Gamma\left(n+\frac{11}{6}\right)}\left|\frac{s_z}{\gamma_z D}\right|^2\right] \tag{80}$$

Expressions used in section 4.4.

When the filter function is equal to

$$F_{n,m}(\kappa,\ z) = [N_n(\kappa) - N_n(\alpha_z\kappa)]^2$$

The integrations can be expressed by

$$I_n(z) = \frac{(n+1)\Gamma\left(n-\frac{5}{6}\right)D^{\frac{5}{3}}}{2^{2n+\frac{11}{3}}\Gamma\left(\frac{17}{6}\right)\Gamma(n+2)}\left\{\frac{8\Gamma\left(\frac{4}{3}\right)\Gamma(2n+3)\left(\alpha_z^{\frac{5}{3}}+1\right)}{3\Gamma\left(n+\frac{3}{2}\right)\Gamma\left(n+\frac{23}{6}\right)} - \right.$$

$$- \frac{2^{2n+1/3}\alpha_z^n}{6n+17}\left[5\left(\alpha_z^2-1\right)\ _2F_1(1/6,\ n-5/6;n+2;\alpha_z^2) - \right.$$

$$\left.\left. - \left(6\alpha_z^2 n - 6n - 22\right)\ _2F_1\left(-5/6,\ n-5/6;n+2;\alpha_z^2\right)\right]\right\}$$

When the beacon is high enough ($H \gg z$), we can obtain the following asymptotic solution expanding to second-order turbulence moments:

$$I_n(z) = \frac{8(n+1)\left[108n(n+2)-55\right]}{6237\ \pi\ \Gamma(-10/3)\ \Gamma(n+23/6)}\Gamma^2\left(-\frac{8}{3}\right)\Gamma\left(n-\frac{5}{6}\right)\left(\frac{z}{H}\right)^2 D^{5/3} \tag{81}$$

Expressions used in section 4.5.

When the filter function is equal to

$$F_{n,m}(\kappa,\ z) = N_n^2(\kappa)[1 - G_s(\kappa,\ z)]^2$$

For the Gaussian source Eq. (12), the results can be expressed as:

$$I_n(z) = \frac{1}{2}(1+n)\Gamma\left(n-\frac{5}{6}\right)\left\{\frac{2^{\frac{1}{3}}\Gamma\left(\frac{7}{3}\right)D^{\frac{5}{3}}}{\Gamma\left(\frac{17}{6}\right)\Gamma\left(n+\frac{23}{6}\right)} + \right.$$

$$+ \frac{(\theta_g z)^{-2n+5/3}D^{2n}}{2^{4n}[\Gamma(n+2)]^2}\left[\ _2F_2\left(n-5/6,\ n+3/2;n+2,\ 2n+3;\ -\frac{D^2}{4(\theta_g z)^2}\right) - \right.$$

$$\left.\left. -2^{n+1/6}\ _2F_2\left(n-5/6,\ n+3/2;n+2,\ 2n+3;\ -\frac{D^2}{2(\theta_g z)^2}\right)\right]\right\}$$

If the widths of the distributed source are very large ($\theta_r z \gg D$), expanding the solutions to second-order terms of small parameter $D/(\theta_r z)$, the results are obtained as follows:

$$I_n(z) = \frac{(1+n)\Gamma\left(n-\frac{5}{6}\right)}{2^{2/3}D^{-5/3}}\left\{\frac{\Gamma\left(\frac{7}{3}\right)/\Gamma\left(\frac{17}{6}\right)}{\sqrt{\pi}\ \Gamma\left(n+\frac{23}{6}\right)} - \frac{(2\theta_r z/D)^{5/3-2n}}{[2^{n+1}\Gamma(n+2)]^2}\left[\left(2^{n+1/6}-1\right) + \frac{\left(n-\frac{5}{6}\right)\left(2^{n+7/6}-1\right)}{2(2+n)(2\theta_r z/D)^2}\right]\right\} \tag{82}$$

Author details

Jingyuan Chen and Xiang Chang

Yunnan Astronomical Observatory, Chinese Academy of Science, China

References

[1] Andrews, L. C. (1998). Special functions of mathematics for engineers. SPIE-International Society for Optical Engineering, Bellingham

[2] Buscher, D. F, Love, G. D, et al. (2002). Laser beacon wave-front sensing without focal anisoplanatism. *Opt. Lett., 27*(3), 149-151.

[3] Dickey, J. O, Bender, P. L, et al. (1994). Lunar laser ranging- A continuing legacy of the Apollo program. *Science 265*(5171), 482-490.

[4] Esposito, S, Ragazzoni, R, et al. (2000). Absolute tilt from a laser guide star: a first experiment. *Experimental Astronomy , 10*(1), 135-145.

[5] Esposito, S, Riccardi, A, et al. (1996). Focus anisoplanatism effects on tip-tilt compensation for adaptive optics with use of a sodium laser beacon as a tracking reference. *J.Opt.Soc.Am.A, , 13*(9), 1916-1923.

[6] Foy, R, Migus, A, et al. (1995). The polychromatic artificial sodium star: A new concept for correcting the atmospheric tilt. *Astron. Astrophys. Suppl. Ser., 111*(3), 569-578.

[7] Fried, D. L. (1982). Anisoplanatism in adaptive optics. *J.Opt.Soc.Am, , 72*(1), 52-61.

[8] Fried, D. L. (1990). Time-delay-induced mean-square error in adaptive optics. *J.Opt.Soc.Am.A , 7*(7), 1224-1225.

[9] Fried, D. L. (1995). Focus anisoplanatism in the limit of infinitely many artificial-guide-star reference spots. *J.Opt.Soc.Am.A, , 12*(5), 939-949.

[10] Gavel, D. T, Morris, J. R, et al. (1994). Systematic design and analysis of laser-guide-star adaptive-optics systems for large telescopes. *J.Opt.Soc.Am.A, , 11*(2), 914-924.

[11] Happer, W, Macdonald, G. J, et al. (1994). Atmospheric-turbulence compensation by resonant optical backscattering from the sodium layer in the upper atmosphere. *J.Opt.Soc.Am.A, , 11*(1), 263-276.

[12] Hardy, J. W. (1998). Adaptive optics for astronomical telescopes. Oxford University Press.

[13] Marc, F, De Chatellus, H. G, et al. (2009). Effects of laser beam propagation and saturation on the spatial shape of sodium laser guide stars. *Optics Express 17*(7), 4920-4931.

[14] Muller, N, Michau, V, et al. (2011). Differential focal anisoplanatism in laser guide star wavefront sensing on extremely large telescopes. *Opt. Lett., 36*(20), 4071-4073.

[15] Neyman, C. R. (1996). Focus anisoplanatism: A limit to the determination of tip-tilt with laser guide stars. *Opt. Lett., 21*(22), 1806-1808.

[16] Parenti, R. R, & Sasiela, R. J. (1994). Laser-guide-star systems for astronomical applications. *J.Opt.Soc.Am.A,* , 11(1), 288-309.

[17] Ragazzoni, R, & Le, B. Roux, et al. ((2005). Multi-Conjugate Adaptive Optics for ELTs: constraints and limitations. *C. R. Phys. 6*(10), 1081-1088.

[18] Riepl, S, Schluter, W, et al. (1999). Evaluation of an SLR adaptive optics system. Laser Radar Ranging and Atmospheric Lidar Techniques II. U. Schreiber and C. Werner. Bellingham. *Proc. SPIE.* , 3865, 90-95.

[19] Rigaut, F, & Gendron, E. (1992). Laser guide star in adaptive optics-The tilt determination problem. *Astron.Astrophys.* , 261(2), 677-684.

[20] Sasiela, R. J. (1992). Strehl ratios with various types of anisoplanatism. *J.Opt.Soc.Am.A,* , 9(8), 1398-1405.

[21] Sasiela, R. J. (1994). Wave-front correction by one or more synthetic beacons. *J.Opt.Soc.Am.A,* , 11(1), 379-393.

[22] Sasiela, R. J. (2007). Electromagnetic wave propagation in turbulence: evaluation and application of Mellin transforms. SPIE Press.

[23] Sasiela, R. J, & Shelton, J. D. (1993). Transverse spectral filtering and Mellin transform techniques applied to the effect of outer scale on tilt and tilt anisoplanatism. *J.Opt.Soc.Am.A,* , 10(4), 646-660.

[24] Stroud, P. D. (1996). Anisoplanatism in adaptive optics compensation of a focused beam with use of distributed beacons. *J.Opt.Soc.Am.A,* , 13(4), 868-874.

[25] Tyler, G. A. (1994). Bandwidth considerations for tracking through turbulence. *J.Opt.Soc.Am.A,* , 11(1), 358-367.

[26] Tyler, G. A. (1994). Rapid evaluation of d0 - the effective diameter of a laser-guide-star adaptive-optics system. *J.Opt.Soc.Am.A,* 11(1): 325-338.

[27] Tyler, G. A. (1994). Wave-front compensation for imaging with off-axis guide stars. *J.Opt.Soc.Am.A,* , 11(1), 339-346.

[28] Tyson, R. (2011). Principles of adaptive optics (3rd Edition ed.), CRC Press, Boca Raton.

[29] Wallner, E. P. (1977). Minimizing atmospheric dispersion effects in compensated imaging. *J.Opt.Soc.Am.,* , 67(3), 407-409.

[30] Welsh, B. M, & Gardner, C. S. (1991). Effects of turbulence-induced anisoplanatism on the imaging performance of adaptive-astronomical telescopes using laser guide stars. *J.Opt.Soc.Am.A,* , 8(1), 69-80.

[31] Whiteley, M. R, Roggemann, M. C, et al. (1998). Temporal properties of the Zernike expansion coefficients of turbulence-induced phase aberrations for aperture and source motion. *J.Opt.Soc.Am.A,* , 15(4), 993-1005.

[32] Whiteley, M. R, Welsh, B. M, et al. (1998). optimal modal wave-front compensation for anisoplanatism in adaptive optics. *J.Opt.Soc.Am.A,* , 15(8), 2097-2106.

[33] Wilson, K. E. (1994). An overview of the Compensated earth-Moon-earth laser link (CEMERLL) experiment. Bellingham. *Proc. SPIE.* , 2123, 66-74.

Permissions

The contributors of this book come from diverse backgrounds, making this book a truly international effort. This book will bring forth new frontiers with its revolutionizing research information and detailed analysis of the nascent developments around the world.

We would like to thank Robert K. Tyson, Ph.D., for lending his expertise to make the book truly unique. He has played a crucial role in the development of this book. Without his invaluable contribution this book wouldn't have been possible. He has made vital efforts to compile up to date information on the varied aspects of this subject to make this book a valuable addition to the collection of many professionals and students.

This book was conceptualized with the vision of imparting up-to-date information and advanced data in this field. To ensure the same, a matchless editorial board was set up. Every individual on the board went through rigorous rounds of assessment to prove their worth. After which they invested a large part of their time researching and compiling the most relevant data for our readers. Conferences and sessions were held from time to time between the editorial board and the contributing authors to present the data in the most comprehensible form. The editorial team has worked tirelessly to provide valuable and valid information to help people across the globe.

Every chapter published in this book has been scrutinized by our experts. Their significance has been extensively debated. The topics covered herein carry significant findings which will fuel the growth of the discipline. They may even be implemented as practical applications or may be referred to as a beginning point for another development. Chapters in this book were first published by InTech; hereby published with permission under the Creative Commons Attribution License or equivalent.

The editorial board has been involved in producing this book since its inception. They have spent rigorous hours researching and exploring the diverse topics which have resulted in the successful publishing of this book. They have passed on their knowledge of decades through this book. To expedite this challenging task, the publisher supported the team at every step. A small team of assistant editors was also appointed to further simplify the editing procedure and attain best results for the readers.

Our editorial team has been hand-picked from every corner of the world. Their multi-ethnicity adds dynamic inputs to the discussions which result in innovative outcomes. These outcomes are then further discussed with the researchers and contributors who give their valuable feedback and opinion regarding the same. The feedback is then collaborated with the researches and they are edited in a comprehensive manner to aid the understanding of the subject.

Apart from the editorial board, the designing team has also invested a significant amount of their time in understanding the subject and creating the most relevant covers. They scrutinized every image to scout for the most suitable representation of the subject and create an appropriate cover for the book.

The publishing team has been involved in this book since its early stages. They were actively engaged in every process, be it collecting the data, connecting with the contributors or procuring relevant information. The team has been an ardent support to the editorial, designing and production team. Their endless efforts to recruit the best for this project, has resulted in the accomplishment of this book. They are a veteran in the field of academics and their pool of knowledge is as vast as their experience in printing. Their expertise and guidance has proved useful at every step. Their uncompromising quality standards have made this book an exceptional effort. Their encouragement from time to time has been an inspiration for everyone.

The publisher and the editorial board hope that this book will prove to be a valuable piece of knowledge for researchers, students, practitioners and scholars across the globe.

List of Contributors

Ren Deqing
Physics & Astronomy Department, California State University Northridge, USA
National Astronomical Observatories/Nanjing Institute of Astronomical Optics & Technology, Chinese Academy of Sciences, China
Key Laboratory of Astronomical Optics & Technology, National Astronomical Observatories, Chinese Academy of Sciences, China

Zhu Yongtian
National Astronomical Observatories/Nanjing Institute of Astronomical Optics & Technology, Chinese Academy of Sciences, China
Key Laboratory of Astronomical Optics & Technology, National Astronomical Observatories, Chinese Academy of Sciences, China

Zoran Popovic, Jörgen Thaung and Per Knutsson
Department of Ophthalmology, University of Gothenburg, Gothenburg, Sweden

Mette Owner-Petersen
Retired from the Telescope Group, Lund University, Lund, Sweden

S. Bonora
CNR-IFN, Laboratory for UV and X-Ray and Optical Research, Padova, Italy

R.J. Zawadzki
VSRI, Department of Ophthalmology and Vision Science, University of California Davis, Sacramento, CA, USA

G. Naletto
CNR-IFN, Laboratory for UV and X-Ray and Optical Research, Padova, Italy
Department of Information Engineering, University of Padova, Padova, Italy

S. Residori
INLN, Université de Nice-Sophia Antipolis, CNRS, France

Thomas Ruppel
Carl Zeiss AG, Corporate Research and Technology, Optronics Systems, Jena, Germany

Li Xuan, Zhaoliang Cao, Quanquan Mu, Lifa Hu and Zenghui Peng
State Key Laboratory of Applied Optics, Changchun Institute of Optics, Fine Mechanics and Physics, Chinese Academy of Sciences, Jilin Changchun, China

Mathieu Aubailly
Intelligent Optics Laboratory, Institute for Systems Research, University of Maryland, College
Park, Maryland, USA

Mikhail A. Vorontsov
Intelligent Optics Laboratory, School of Engineering, University of Dayton, Dayton, Ohio, USA

S. G. Garanin and F. A. Starikov
Russian Federal Nuclear Center –VNIIEF, Institute of Laser Physics Research, Russia

Yu. I. Malakhov
International Science and Technology Center, Russia

Jingyuan Chen and Xiang Chang
Yunnan Astronomical Observatory, Chinese Academy of Science, China